THE WARDENCLYFFE BLUEPRINTS

By Ernst Willem van den Bergh

ii

(this page is intentionally left blank)
(this page is intentionally left blank)

Contents

Preface

Through the help of my friend Tibor Hrs Pandur, I have found where the original blueprints of the Wardenclyffe Tower were stored; The Nikola Tesla Museum in Belgrade. Who would have guessed?

I contacted the museum and was allowed access to photos of these blueprints so that I could make and sell replicas of these blueprints. I don't want to jeopardize my good relationship with the museum by giving away what they charge 3 EUR per page for. That is why I will only show a small thumbnail of the photo from the museum, then I am free to show what I have made of it. For those who want access to the originals, you can contact the museum (info@tesla-museum.org), answer a couple of questions, send a copy of your passport and pay 3 EUR per page to get access. (this, of course, goes for all documents mentioned in this book)

I will include the museum document numbers so you know what to ask for.

What you'll get is a photo – not a scan – of the originals, which were folded in many ways, some are even damaged, most are discoloured and, of course, the photos are not to scale.

My replicas solve those issues. I have gone to great lengths to get the dimensions exactly right. So if you want a full or half-size replica, you can contact me. You'll find details on how to order my replicas through the QR codes at the end of this book.

I want to thank my friend Kyle Dell'Aquila for the renderings of his Wardenclyffe model that he made in Blender. Kyle has made the most accurate model that I think exists today, together with breathtaking models of how it may have looked when completed. These renderings bring the past to life.

I'm a man of few words, and for that reason I feel especially honoured that Marc J. Seifer has agreed to write an introduction to this book.

And, of course, I want to thank my friend Tibor who has helped my research enormously by sharing thousands of museum documents with me. A massive thank you to the three of you!

Ernst.

Introduction

By Marc J. Seifer, Ph.D.
© June 5, 2024

The year was 1976 and I had begun a career as a journalist, writing for several scientific magazines on consciousness including a rather exotic magazine *Ancient Astronauts*, which speculated that extraterrestrials had indeed visited the planet Earth. At that time, due to Nikola Tesla's hypothesis that he had, in fact, received intelligent impulses from a nearby planet, most likely Mars, Tesla got lumped into the UFO crowd and that further helped cause his name to disappear from the mainstream, kept alive in this more esoteric realm.

On assignment on a different topic, I chanced upon Tesla's name in a book on Avatars, enlightened beings who descended to our planet to help Earthlings deal with the present-day crisis, which at that time, the mid-late 1800's, was air pollution caused by the numerous coal-operated local power plants which were springing up at every urban center for the specific task of providing electricity for lighting homes. Because the present-day system of direct current was so inefficient, these power plants could only transmit electricity about one mile, power dropping off over distance. This means that if your home was close to the power source, it would be brightly lit, and if the home was farther away, your lightbulbs would be dimmer.

The reason why Tesla appeared in this occult text was because his system of figuring out how to harness alternating current led directly to his invention of what became the hydroelectric power system, a completely pollution-free invention that made all of the thousands of local coal-operated smoke-spewing power plants obsolete. Tesla's invention, put in at Niagara Falls by George Westinghouse created a new invention that became the backbone of our modern electronic age. Because the invention, which encompassed the creation of the induction motor, was fundamental, it is basically unchanged even today well over a century later.

My research began by reading John O'Neill's classic 1944 Tesla biography *Prodigal Genius*. However, if I had one criticism of this truly terrific biography, is that it had no photographs. This led me to Hunt and Draper's 1964 book *Lightning in His Hand: The Life Story of Tesla* which I found at the University of Rhode Island's superb library where I was doing most of my research, now working on a doctoral dissertation trying to figure out all the reasons why this great inventor's name had slipped into total obscurity.

On the inside front and black flap of this excellent book I saw the following photograph for the first time in my life:

Nikola Tesla in a double exposure photo taken at Colorado Spr.
Nikola Tesla in a double exposure photo taken at Colorado Springs
in 1899 by Dickenson Alley, *Seifer Archives*.

When I saw this photo, I had the strangest feeling of déjà vu even though I don't think I had ever seen the picture before. However, there was just something about it that resonated deep inside of me.

As my research continued, I realized that one of the key reasons why Tesla's name disappeared from the history books was because he failed at Wardenclyffe, his great wireless enterprise which he undertook out on Long Island about 60 miles from the heart of New York City, and the Waldorf Astoria hotel where Tesla was living.

I then set out to figure out why Wardenclyffe failed and this took me to the microfilm letters that the Library of Congress had on Tesla's correspondence Tesla had with J. Pierpont Morgan, the great financier who had funded this operation. Essentially, what I discovered was that because Tesla felt that Guglielmo Marconi was pirating his apparatus, Tesla constructed a tower literally twice the height that he had promised to build with the $150,000 Morgan had given him.

THE ELECTRICAL EXPERIMENTER March, 1916

The Tesla High Frequency Oscillator

By H. Winfield Secor, E. E.

THE ELECTRICAL EXPERIMENTER March, 1916

The Tesla High Frequency Oscillator

By H. Winfield Secor, E. E.

Fig. 1. Probably Appearing as the Most Wonderful Earth Oscillator Tower and Night-Field Illuminated at Shoreham, L. I. and Is Intended for Radiating Electrical Energy in the Form of High Frequency Waves Dissipated Thru the Earth Itself. Tower Stands 187 ft. High.

A spectacular drawing from Hugo Gernsback's magazine Electrical Experimenter which shows what Wardenclyffe would supposedly look like when lit at night, *Seifer Archives.*

Tesla argued that even though the costs might double, since the tower would now not just send messages to Europe, but also to South America, Africa, Asia and Australia as well, revenues would be at least five times greater than merely double. However, Morgan saw this as a breach of contract and all of this happened in 1901 when Marconi, with much inferior equipment, succeeded in sending the Morse code for the letter "S"; dot-dot-dot, across the Atlantic, thus beating Tesla in the race to send the first wireless message across that mighty ocean.

And so Tesla's great operation at Wardenclyffe lay idle. Due to his inability to continue to pay his rent at the Waldorf, Tesla owed them nearly $20,000; he turned the property over to the hotel and they, in turn in 1917; tore the 187-foot tower down and sold the parts for salvage.

As I continued my research which resulted in the book *Wizard: The Life and Times of Nikola Tesla: Biography of a Genius*, which was first published in 1996, I also spent a considerable amount of time trying to figure out if indeed Wardenclyffe would have been viable had it been completed.

A big difference between Tesla's system and that of Marconi's was that for nearly 20 years, all Marconi could do was transmit Morse code. His technology was rather primitive. He could not really create separate channels and he did not have the technology to transmit voice or pictures. Tesla's patents, on the other hand, were very clear in explaining that his technology was indeed set up to transmit voice and thus music, as well as pictures, and as Tesla claimed, also power.

It took me many years to realize that the amazing 1899 photo of Tesla surrounded by lightning in his Colorado Springs laboratory, which was an experimental station for sending wireless messages, was created for one reason only, PR, public relations. The drawing from 1916 in *Electrical Experimenter* with energy radiating off the top of Tesla's tower was also plain old hype. That glow above the tower's cupola had no relationship whatever to Tesla's actual plans as to how the tower would operate.

In 2017, I received a call from Hollywood, Prometheus Films, asking me if I wanted to star in a television show called *The Tesla Files*. The plot would be to try and figure out all of Tesla's secrets which had been hidden in his many trunks which he had in a warehouse in New York City where they were found after he died during the height of WWII in January of 1943. For various reasons, the U.S. government retained possession of these trunks for nearly a decade. And then, in the early 1950's most, if not all of them were shipped to Belgrade so that Tesla's nephew, Sava Kosanovic, who was also the ambassador from Yugoslavia, would create a museum there in Tesla's honor.

There were several questions. Did the U.S. military retain any records that were not shipped to Belgrade? Did the U.S. military use any of this information to create secret weapons or other technologies? Had they, for instance, figured out how to transmit electrical power according to some secret plan that Tesla never completed?

Although the Waldorf Astoria had been given Wardenclyffe as collateral against the $20,000 Tesla owed in back rent, after they destroyed the tower, they only received several thousand dollars in salvage and in selling the property and so they sued Tesla to recoup the balance which would have been about $15,000.

Tesla therefore testified in several court cases and during this time he revealed certain aspects of the tower that had never been revealed before. Tesla argued that the equipment he had put in the care of the Waldorf was worth well over $60,000. From Tesla's point of view, had he been given the funds to complete the tower he would have had a world-wide wireless telephone system, which, as I explained in *Wizard* "was [my] primary purpose of the tower, your Honor... to telephone, send the human voice and likeness [photographs] around the globe" (p. 408). In other words, the tower was worth potentially millions of dollars and the Waldorf, instead of realizing the great gem that they owned, destroyed it for a pittance.

When we started shooting *The Tesla Files*, which starred myself, investigator Jason Stapleton, ILM editor Tim Eaton who was my writing partner on the Tesla film we are working on and the brilliant physicist Travis Taylor, Ph.D., one of our primary goals was to locate Tesla's blueprints. Another goal was to figure out precisely the real meaning of Tesla's additional testimony concerning what happened beneath the ground at Wardenclyffe.

Tesla said it this way: "You see, the underground work was one of the most expensive parts of the tower." Tesla was referring specifically to special apparatus he invented for "gripping the earth."

"The shaft, your Honor, was first covered with timber and the inside with steel. In the center of this there was winding stairs going down and in the center of the stairs was a big shaft again through which the current was to pass, and this shaft was so figured in order to tell exactly where the nodal point is, so that I could calculate exactly the size of the Earth or the diameter of the Earth and measure it exactly within four feet with that machine.

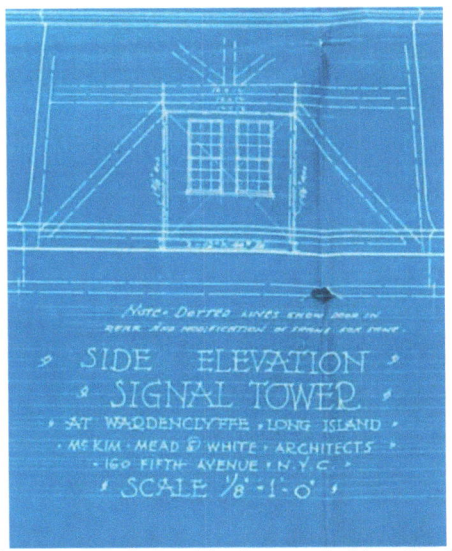

Tesla's tower was designed by the great Architect Stanford White of the well-known firm McKim, Mead & White. The actual construction of the 187' Transmission tower was carried out by W.D. Crow, *Seifer Archives.*

"And then the real expensive work was to connect that central part with the Earth, and there I had special machines rigged up which would push the iron pipes, one length after another, and I pushed, I think sixteen of them, three hundred feet. The current through these pipes [was to] take hold of the Earth. Now that was a very expensive part of the work, but it does not show on the tower, but it belongs to the tower."

At this point, Tesla states, as explained above, that "the primary purpose of the tower" was a global wireless telephone system also able send pictures and even electrical power (*Wizard*, pp. 406-408). In fact, Tesla is the primary inventor of the wireless telephone including the ability to create an unlimited number of wireless channels, as he told Morgan during their negotiations. Tesla achieved what he called "protected privacy," for instance, the ability to create separate radio stations and encrypted transmissions such as when he directed his remote-controlled robotic boat, by multiplying frequencies. This technology was invented by Nikola Tesla and is the reason why every person on the Earth, numbering in the billions, can each have their own separate cell phone.

When we were out at Wardenclyffe, I went to the local courthouse and found Tesla's contract to purchase the land and also a reference to "blueprints." Since Stanford White's home was quite close by, about 15 miles west towards New York City, we sent Jason Stapleton there to see if he could locate McKim, Mead & White's drawings of the tower and we also went to Columbia University to search through the Stanford White papers to no avail.

I also walked the property to locate a small igloo-shaped construction that I had seen back in the 1980's when I first came to Wardenclyffe. This was most likely an air shaft located across Tesla Street by what is now a fire station. Unfortunately, this all-important little brick construction was taken down probably in 2002, shortly after the 911 attack of the Twin Towers in New York City, because at that very location by the fire station, the town built a small memorial to honor the 3,000 people who died when the Wall Street towers collapsed, September 11, 2001.

In June of 2017, we flew to Belgrade Serbia and looked through many documents, some related to Tesla's work at Colorado Springs and some linked to his so-called particle beam weapon and negotiations with the Soviet Union. We also inquired about whether or not Tesla had blueprints for the tower but we continued to hit a dead end.

At Wardenclyffe, because the interior of the building was still polluted, we sent a drone into the laboratory to search the premises. At the same time the producer for Prometheus Films, Kevin Burns hired Hager Geoscience to conduct ground penetrating radar over the property. And I continued to search whatever files I could locate to see if Tesla wrote anything else about the underground portion of the facility.

In 1915, Tesla began working for Telefunken, the German wireless concern through their American company Atlantic Communications. As their top consultant whose patents were the basis of their wireless enterprise, Tesla went out to their wireless plant at Sayville, Long Island, which was located quite near Wardenclyffe, and the first thing Tesla told Jonathan Zenneck, their wireless expert in from Germany, was to increase their ground connections.

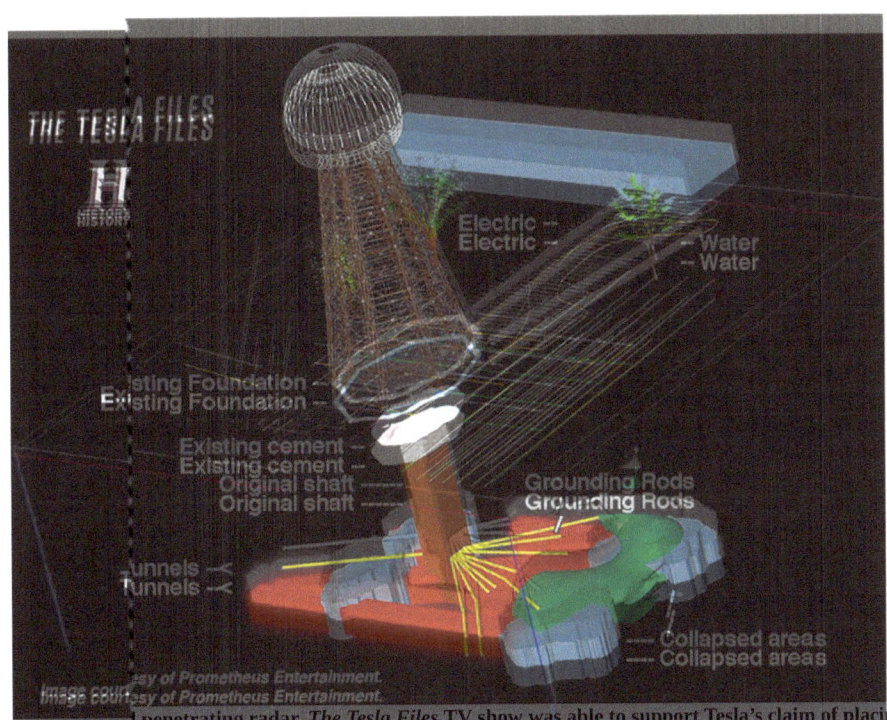

Through ground penetrating radar, *The Tesla Files* TV show was able to support Tesla's claim of placing sixteen "earth grippers" radiating out for a total distance of 300' of tubing to augment the transmission of wireless impulses at his Wardenclyffe lab, *Seifer Archives*.

After listening to Tesla, as I state in *Tesla: Wizard at War* (p. 334), the *New York Times* reported on April 23, 1915, "Germans Triple Wireless Plant."

Shortly thereafter, in a 1916 deposition which discusses Tesla's meeting with Zenneck out at Sayville, Tesla stated that "Wardenclyffe was set up to reduce considerably the radiation aspect and increase the conductive aspect properties by sending EM energy through the Earth." Tesla testified that, "By proper design and choice of wave lengths, you can arrange it so that you get, for instance 5% in these EM waves and 95% in the current that goes through the Earth. That is what I am doing."

Going back to the first two images, the photo of Tesla at Colorado Springs and the drawing of Tesla's tower with the radiation coming off the top of the cupola, after many years of study, I realized that Tesla's goal was not to send wireless messages to radiate off the top of the tower. Rather, the function of the cupola is to collect the electrical energy in various glass bulbs and drive the energy down into the planet, extending the energy out through Tesla's earth grippers, now seen for the first time (above) through the work we did in *The Tesla Files*.

Tesla's Wardenclyffe letterhead created circa 1906 *Seifer Archives.*

Although Marconi also had a ground connection, Marconi and all that followed him were under the impression that the best way to transmit wireless messages was through the airwaves. In fact, that is not at all what Tesla's plan was. He states that he could transmit 95% of the energy through the Earth and lose only 5% by directing all the energy into the planet, not by sending it through the air.

It is my belief that the April 23, 1915 article in the *New York Times* verifying that the Germans tripled their power after increasing their ground connections is one of the most important articles regarding the viability of Wardenclyffe because it verified exactly what Tesla said. He told Zenneck to increase the ground connections and suddenly Telefunken became the most powerful wireless plant on the planet.

I would like to thank Tibor Hrs Pandur for his incredible ability to locate documents that had been unavailable to the public, particularly the Tesla Wardenclyffe blueprints, and also thank Ernst Willem for his tireless work on this topic and for asking me to write an introduction to the Wardenclyffe Blueprints.

The three of us have been working hard to find additional writings that Tesla created concerning his earth grippers and four tunnels which lie about 60 feet below the ground at the site of the Wardenclyffe tower, but to date, no such writings, other than Tesla's testimony referred to above, have been discovered.

Marc J. Seifer, Ph.D., is the author of more than 100 articles and a dozen books, including the acclaimed *Wizard: The Life & Times of Nikola Tesla, Ozone Therapy for the Treatment of Viruses* and his latest *Tesla: Wizard at War.* Having lectured at every

International Tesla Conference held in Colorado Springs from 1984 to 1996, Dr. Seifer has also spoken at Brandeis University, Federal Reserve Bank in Boston, LucasFilms Industrial Light & Magic, at both Oxford and Cambridge Universities in England, West Point Military Academy, the New York Public Library and the United Nations. Featured in *The Washington Post, Wall Street Journal, Scientific American, MIT's Technology Review* and *New York Times*, Marc has appeared on *Coast to Coast radio,* the BBC, NPR's *All Things Considered*, and in the 5-part limited series *The Tesla Files* which he helped create, which has gone out to 40 countries and played on the History Channel.

Map of Wardenclyffe

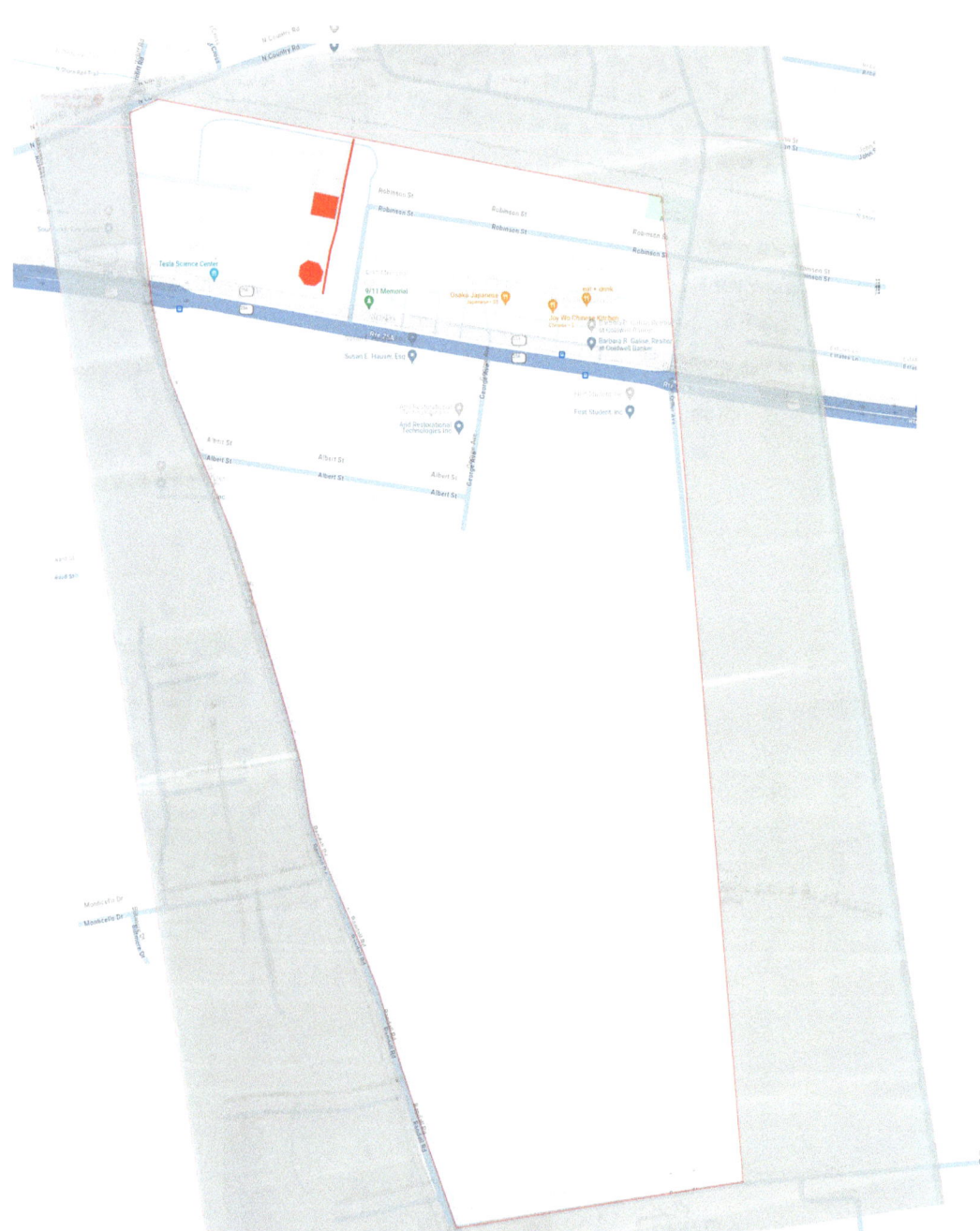

What you see here is a Google map overlaid with the map of Wardenclyffe found in the Tesla museum[1]. The red square is the location of the laboratory, and the octagon south of that is the location where the tower once stood. The red line to the east of those marks the railway that was to bring coal for the boiler in the laboratory.

---

1 This is document number MNT,V,I-4,49.

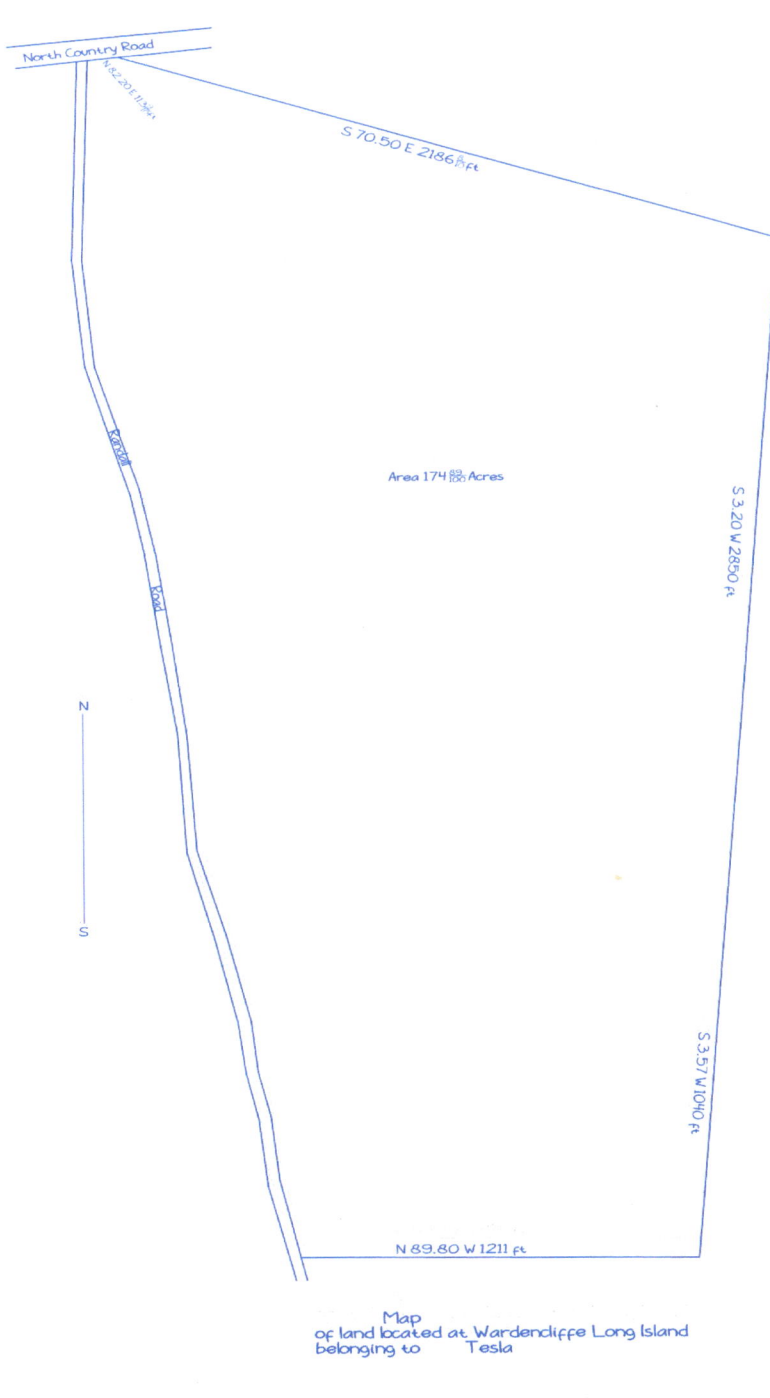

Map
of land located at Wardencliffe Long Island
belonging to   Tesla

Scale 200ft to an inch

It is clear that Tesla had big plans for Wardenclyffe. It was to become the world's communication centre, and more than that... *the next years will see the establishment of an "air-power", and its centre may be not far from New York.*[2]

It is clear that Tesla had big plans for Wardenclyffe. It was to become the world's communication centre, and more than that...

---

2   Quote from "the Problem of Increasing Human Energy", by N. Tesla, June 1900. Could "air-power" refer to power from the air, a means of increasing human energy? (see my book "the Tesla Code")

Tower Side Elevation

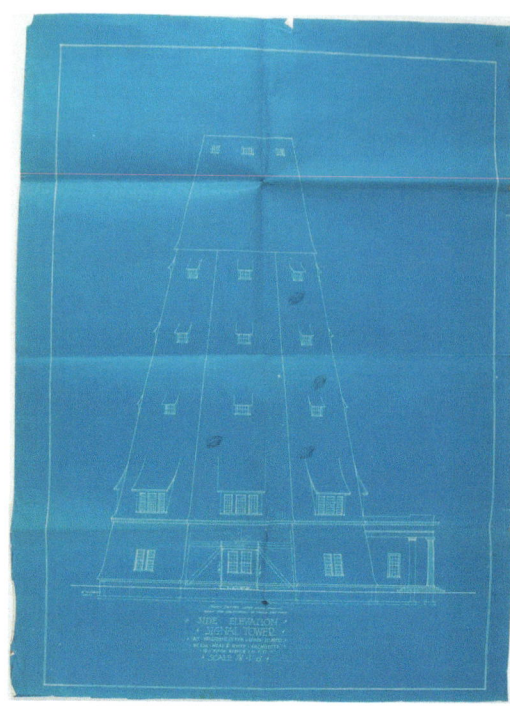

For those who are unfamiliar with architectural terminology, "side elevation" means something like "view from the side of the structure without perspective".

As we normally see everything with perspective this requires some getting used to.

In more mathematical wording, it is a perpendicular projection on a vertical plane next to, and facing the building.

While this is a view from the side of the tower, on the ground level in dotted lines an alternative frame is drawn. According to the text, this alternative frame is to go on the back entrance of the tower to make it possible to put a door there. From the pictures of the tower we can see that this construction is used only at the side away from the laboratory, which would be the front of the tower.

NOTE: DOTTED LINES SHOW DOOR IN
GEAR AND MODIFICATION OF FRAME FOR SAME.

SIDE ELEVATION
SIGNAL TOWER
AT WARDENCLYFFE · LONG ISLAND
· McKIM · MEAD & WHITE · ARCHITECTS ·
· 160 FIFTH AVENUE · N.Y.C ·
SCALE ⅛" · 1'·0'

Tower Front Elevation
Tower Front Elevation

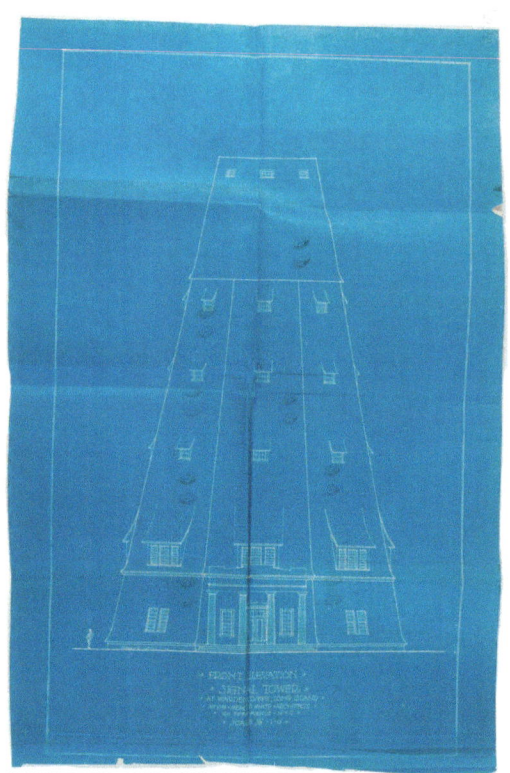

These first two blueprints do not show any measurements, but it has been made to scale such that $\frac{1}{8}$ inch on paper is 1 foot in real life.

What is remarkable is that the windows(?) at the top are inside the cupola as demonstrated by overlaying these images.

As the platform rests on top of this structure, one can easily see that the cupola should be a little higher than depicted. Yet, not high enough to uncover these windows:

GROUND LEVEL

GROUND LEVEL

FRONT ELEVATION
SIGNAL TOWER

SCALE ⅛" · 1'· 0"

16

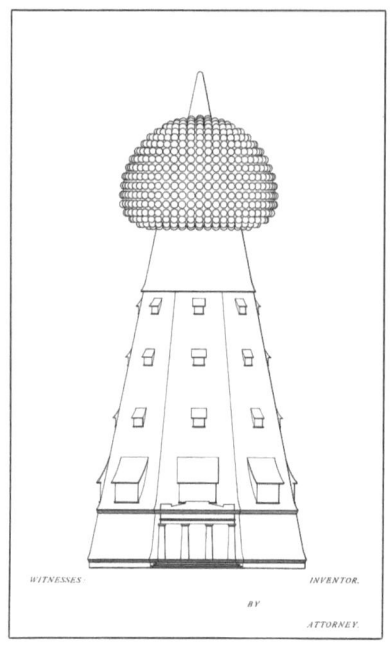

When discussing appearances, we need to consider these images as well.

A peculiar conical object is protruding from the top of the cupola, which is not visible in the other images. This is what Tesla called the "safety valve". When the electrical potential gets too high, it can be raised, creating a break-out point for the electric charge.

But under normal operation, this "valve" would remain hidden inside the cupola.

The concept of a safety valve is also mentioned in patent 1,119,732.

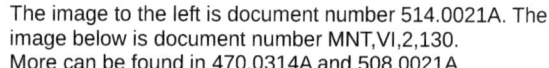

The image to the left is document number 514.0021A. The image below is document number MNT,VI,2,130. More can be found in 470.0314A and 508.0021A

Tower Foundations

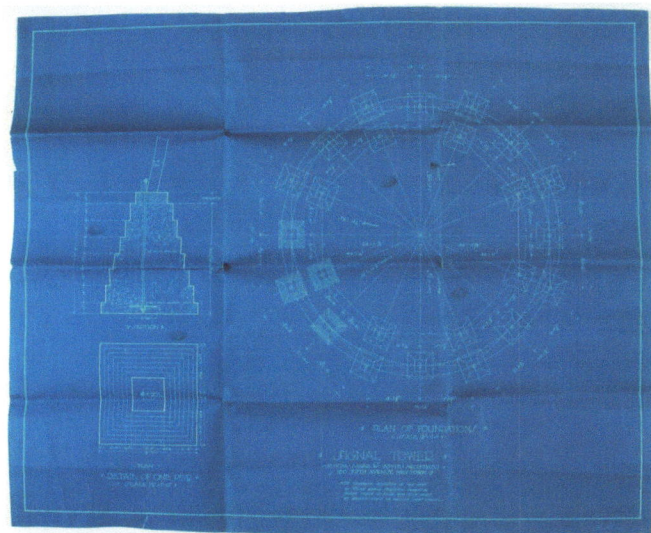

Those who have visited Wardenclyffe have probably seen the granite blocks that once formed the outer ring of the foundation of the tower. (the inner ring has been destroyed)

That piece of granite hides a massive piece of concrete. And a 2¼" (5.7 cm) iron rod through the middle connected this to the wooden frame of the tower.

images by Kyle Dell'Aquila

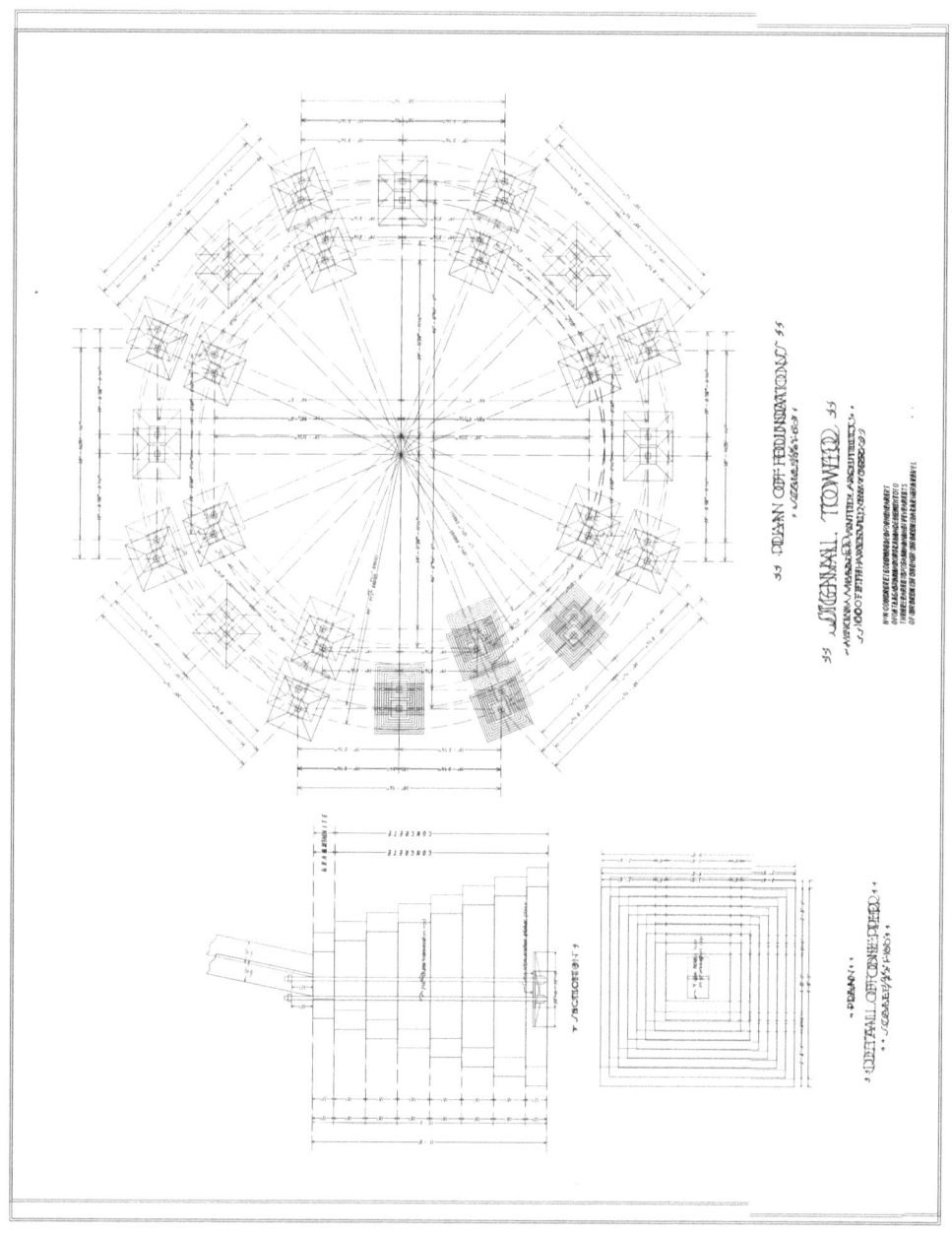

PLAN OFF FOUNDATIONS
SCALE 1/8" = 1'-0"

SIGNAL TOWER
MORGAN MEAD & HOWARD ARCHITECTS
100 FIFTH AVENUE NEW YORK

DETAIL OF ONE PIER
SCALE 1/2" = 1'-0"

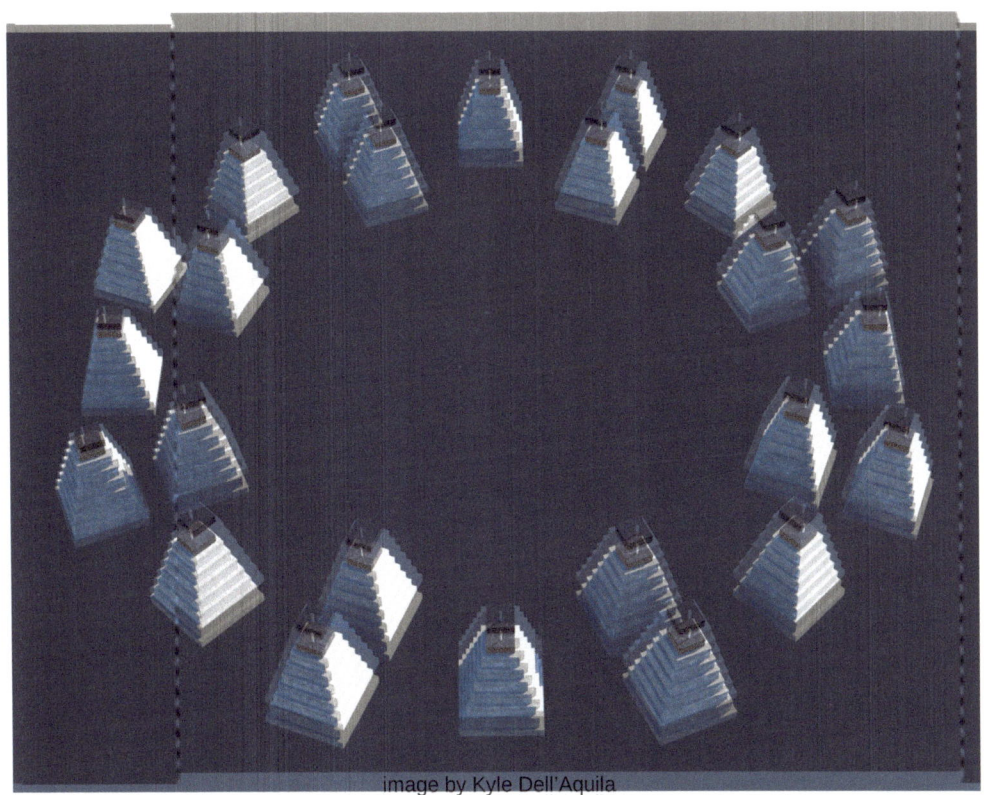

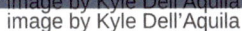
image by Kyle Dell'Aquila
image by Kyle Dell'Aquila

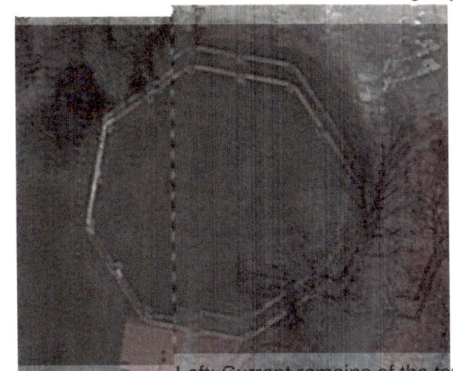 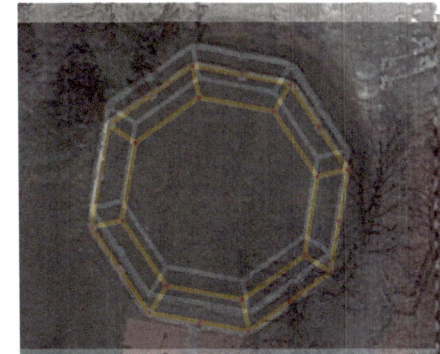

Left: Current remains of the tower foundation. Right: original situation.
Left: Current remains of the tower foundation. Right: original situation.

Tower Frame

The construction of the frame is shown half as "elevation" and half as "section". This blueprint has an error. If you add up the separate heights of each level, you will get 46.8503 meters whereas the total height is given as 154' 2½" (47.0027 m), a difference of 6" (15.24 cm).

From this blueprint alone, it is unclear why there is a line drawn 3' above the top of the posts. I guess that this is the top of the platform, as that works out really well with the other blueprints.

Height			Level	Accumulative height
feet	inch	meter		
15	6	4.7244	B	4.7244
14	6	4.4196	C	9.1440
14	0	4.2672	D	13.4112
13	0	3.9624	E	17.3736
12	0	3.6576	F	21.0312
12	0	3.6576	G	24.6888
11	0	3.3528	H	28.0416
10	0	3.0480	I	31.0896
9	6	2.8956	K	33.9852
9	0	2.7432	L	36.7284
13	6	4.1148	M	40.8432
13	3	4.0386	N	44.8818
3	5.5	1.0541	Top of posts	45.9359
3	0	0.9144	line	46.8503

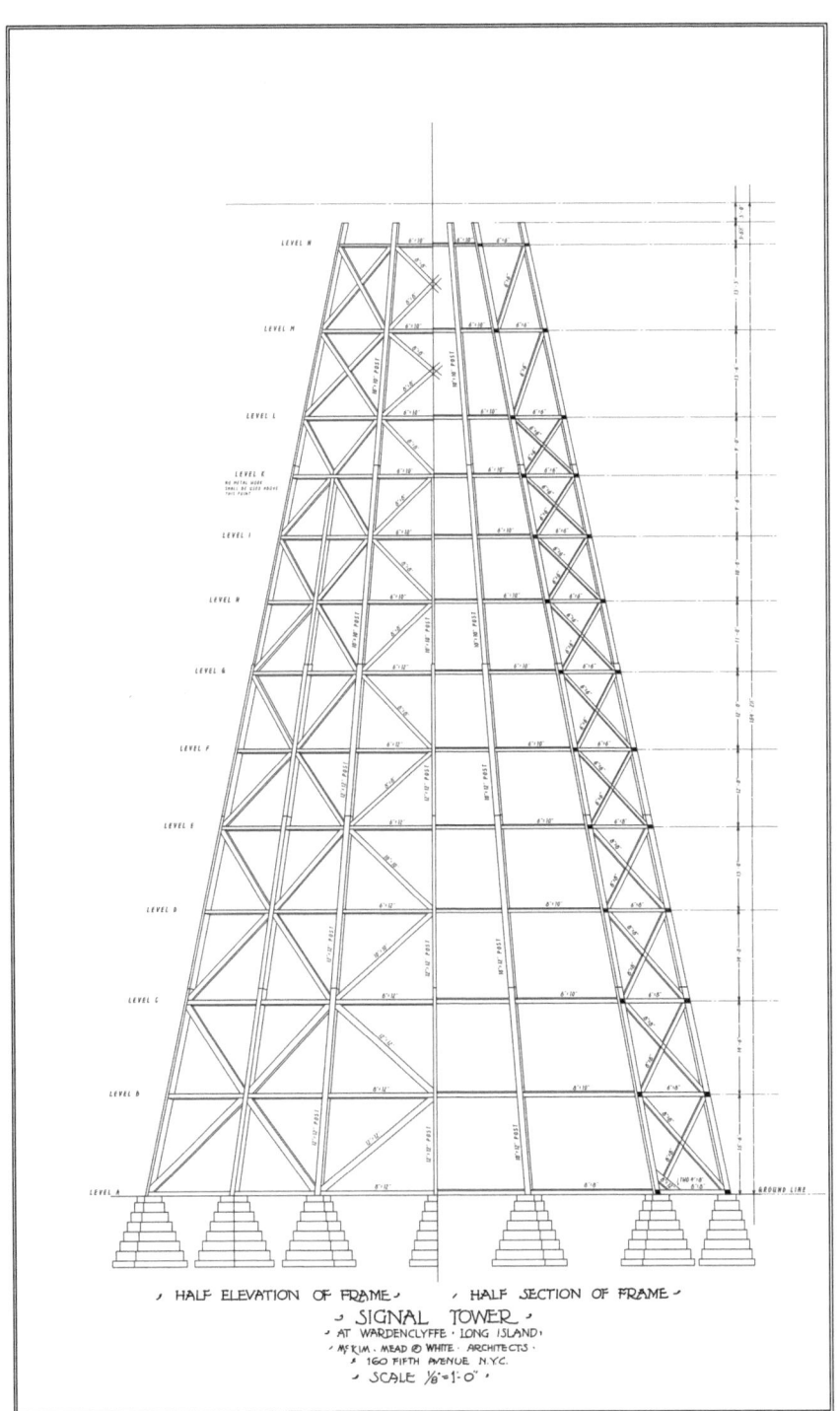

· HALF ELEVATION OF FRAME ·   · HALF SECTION OF FRAME ·

· SIGNAL TOWER ·
· AT WARDENCLYFFE · LONG ISLAND ·
· McKIM · MEAD & WHITE · ARCHITECTS ·
· 160 FIFTH AVENUE N.Y.C. ·
· SCALE ⅛" = 1'- 0" ·

Now, if we take the "half elevation of frame" and mirror it to get a complete elevation of the frame, apply the modification shown in the side elevation and then overlay that with the front elevation of the tower, we get this:

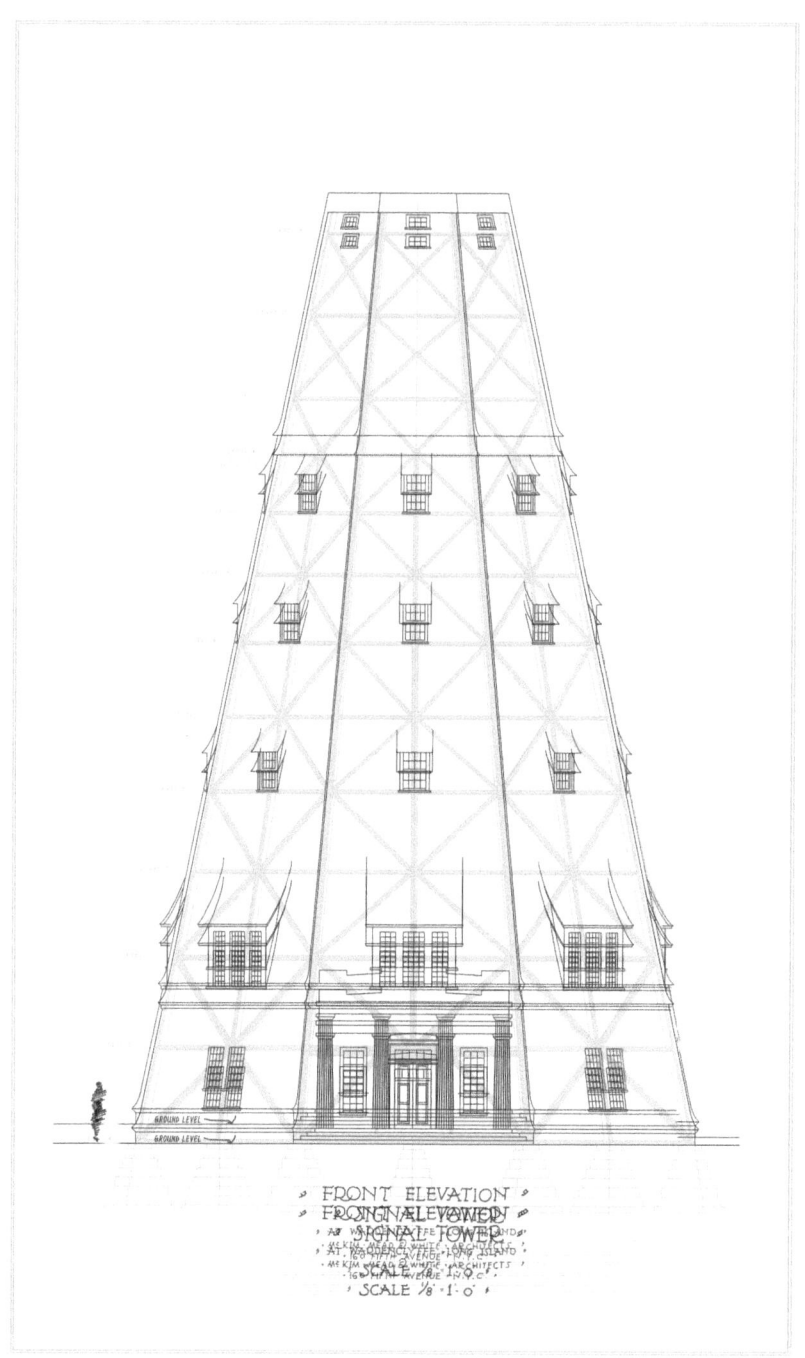

FRONT ELEVATION
SIGNAL TOWER

I think the windows above level B are only there to let some light in and perhaps more for aesthetics than for anything else.

Tower Frame Half Section

Tower Frame Half Section

The tower was to have 2 stairs, one in the front and one in the back.

Here we see how these were constructed on each level except for the ground floor.

The construction of the stairs as seen in the pictures that we have, is very different and located on a different side of the tower, as shown here:

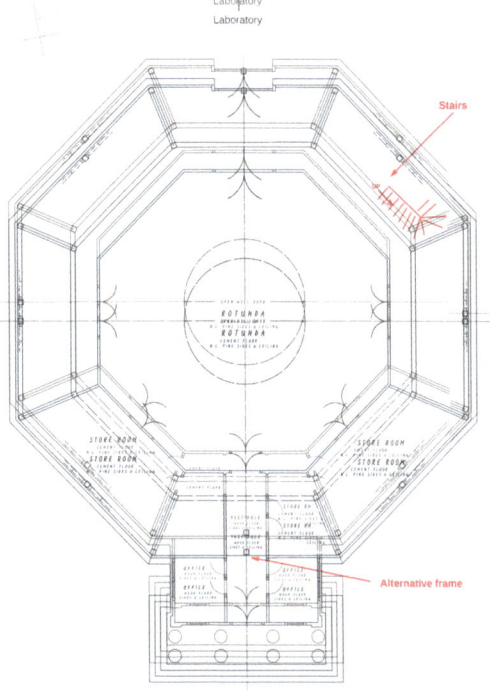

We also get a glimpse of the inside of the tower.

We see "NC pine" being used quite a lot. I think the NC stands for North Carolina here. People outside of the US will probably not immediately know.

Looks like Tesla liked to use this as a finish on the inside.

Here is what an NC pine finish could look like.

The tower was to have 2 stairs, one in the front and one in the back.

Here we see how these were constructed on each level except for the ground floor.

The construction of the stairs as seen in the pictures that we have, is very different and located on a different side of the tower, as shown here:

We also get a glimpse of the inside of the tower.

We see "NC pine" being used quite a lot. I think the NC stands for North Carolina here. People outside of the US will probably not immediately know.

Looks like Tesla liked to use this as a finish on the inside.

Here is what an NC pine finish could look like.

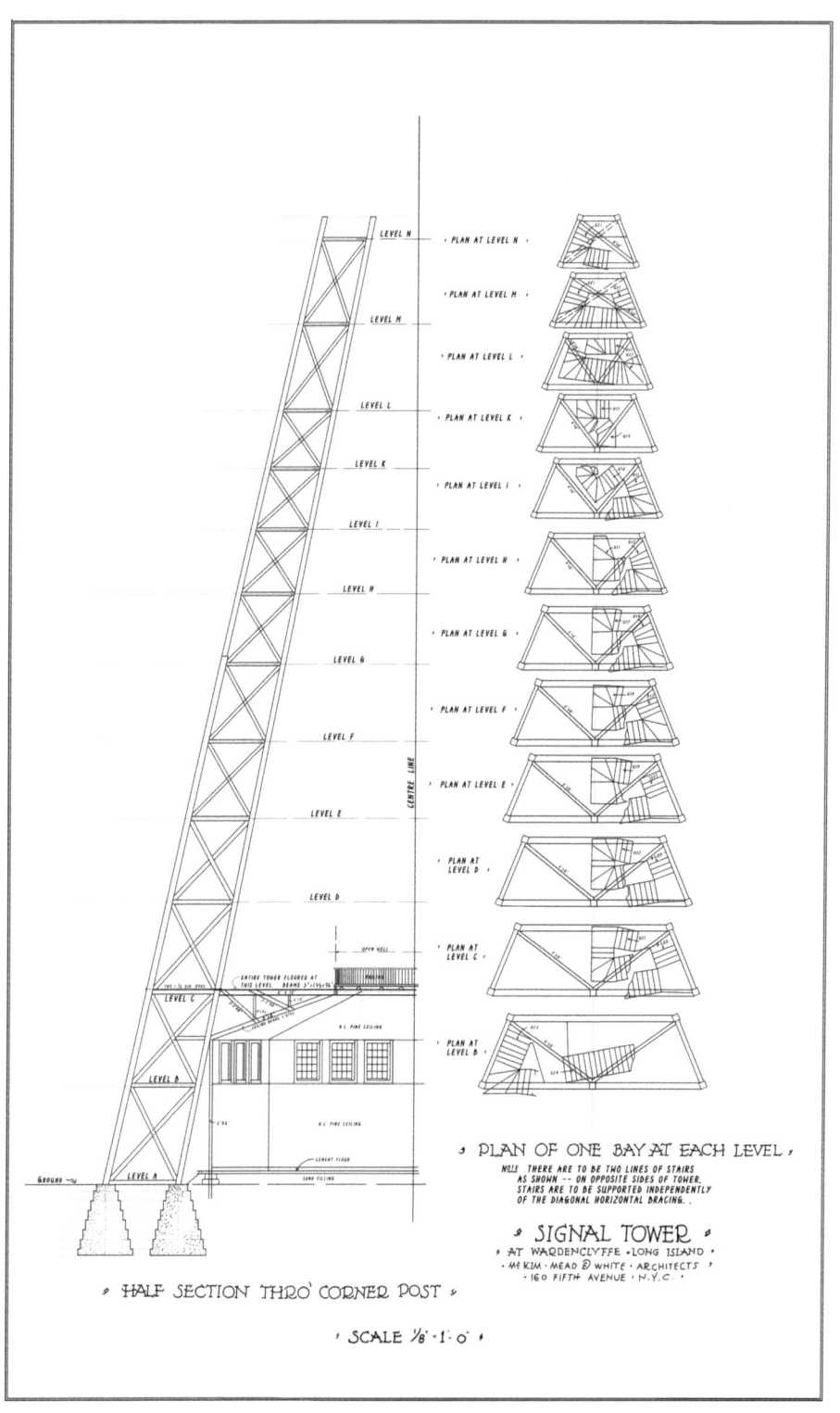

PLAN OF ONE BAY AT EACH LEVEL

NOTE. THERE ARE TO BE TWO LINES OF STAIRS
AS SHOWN -- ON OPPOSITE SIDES OF TOWER.
STAIRS ARE TO BE SUPPORTED INDEPENDENTLY
OF THE DIAGONAL HORIZONTAL BRACING.

SIGNAL TOWER

AT WARDENCLYFFE · LONG ISLAND
Mc KIM · MEAD & WHITE · ARCHITECTS
· 160 FIFTH AVENUE · N.Y.C. ·

HALF SECTION THRO' CORNER POST

SCALE ⅛" · 1'· 0"

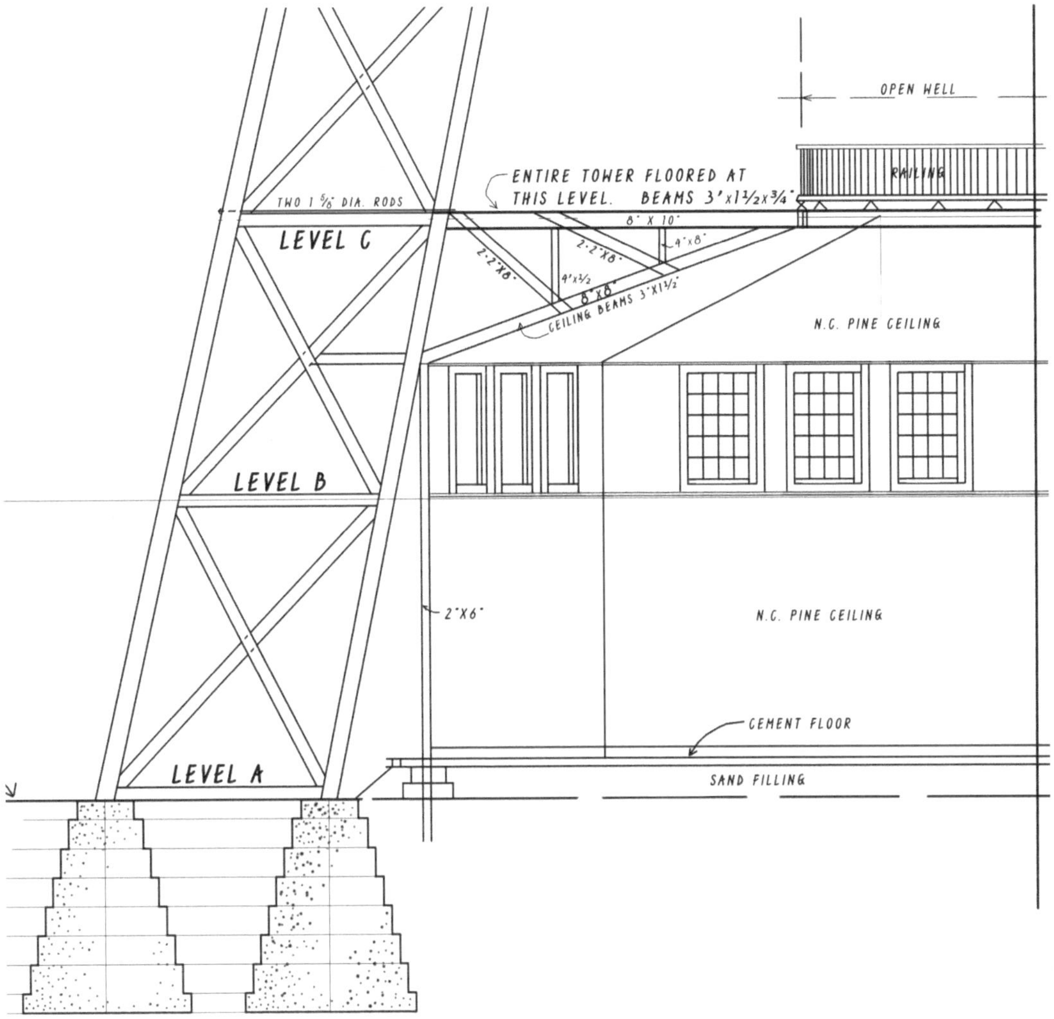

OPEN WELL

RAILING

ENTIRE TOWER FLOORED AT
THIS LEVEL.   BEAMS 3"x1½x¾"

TWO 1 ⅝" DIA. RODS

LEVEL C

8" X 10"

2"-2"X6"

4"x8"

2"-2"X6"

4"x1½

6"X6"

CEILING BEAMS 3"X1½

N.C. PINE CEILING

LEVEL B

N.C. PINE CEILING

2"X6"

CEMENT FLOOR

LEVEL A

SAND FILLING

, PLAN AT LEVEL N ,

, PLAN AT LEVEL M ,

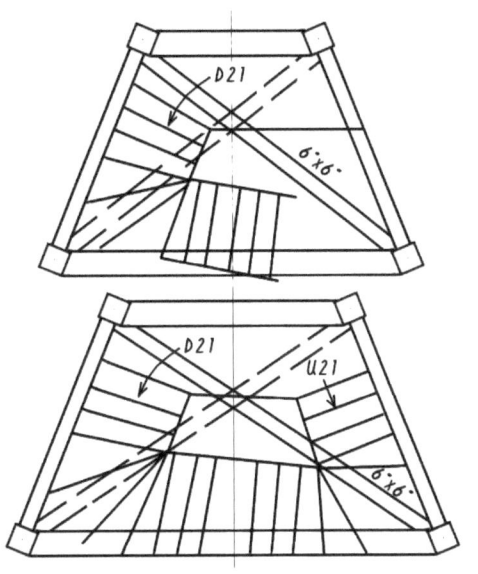

D21

6"x6"

D21

U21

6"x6"

; PLAN AT LEVEL L ;

; PLAN AT LEVEL K ;

; PLAN AT LEVEL I ;

; PLAN AT LEVEL H ;

; PLAN AT LEVEL G ;

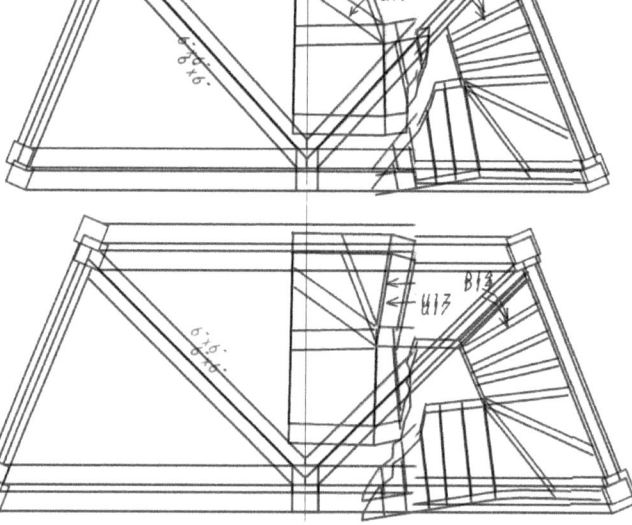

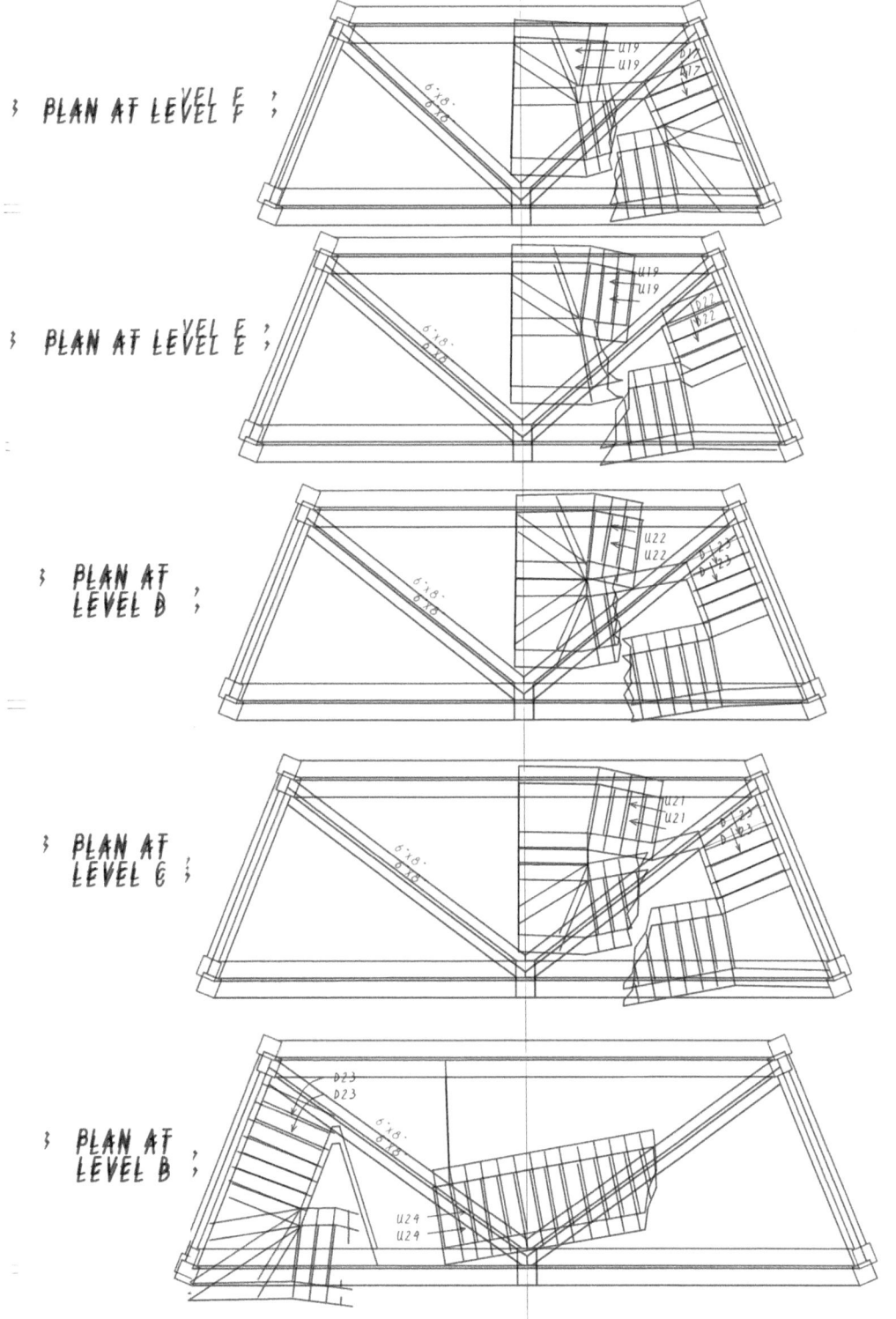

; PLAN AT LEVEL F ;

; PLAN AT LEVEL E ;

; PLAN AT
LEVEL D ;

; PLAN AT
LEVEL C ;

; PLAN AT
LEVEL B ;

image by Kyle Dell'Aquila

Tower Connections

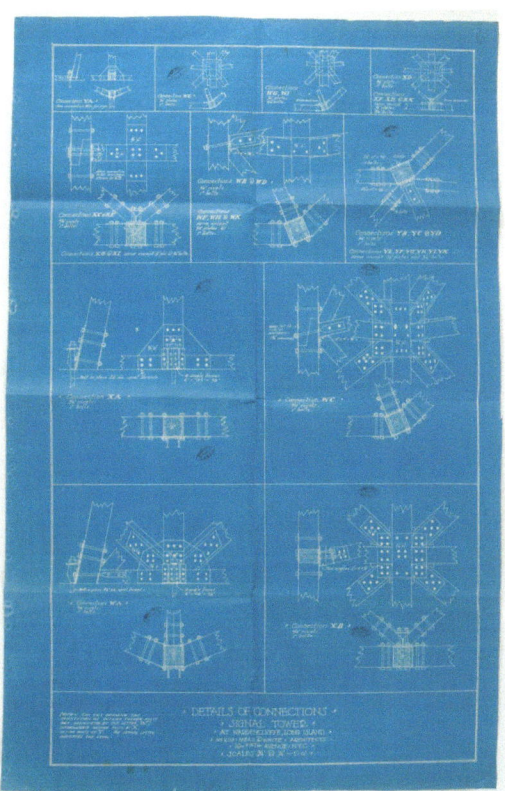

Up to level K, these iron plates were used to connect the woodwork. On the higher levels, I guess, wooden pegs were used. But it is not specified in these blueprints how that was done.

images by Kyle Dell'Aquila

 34

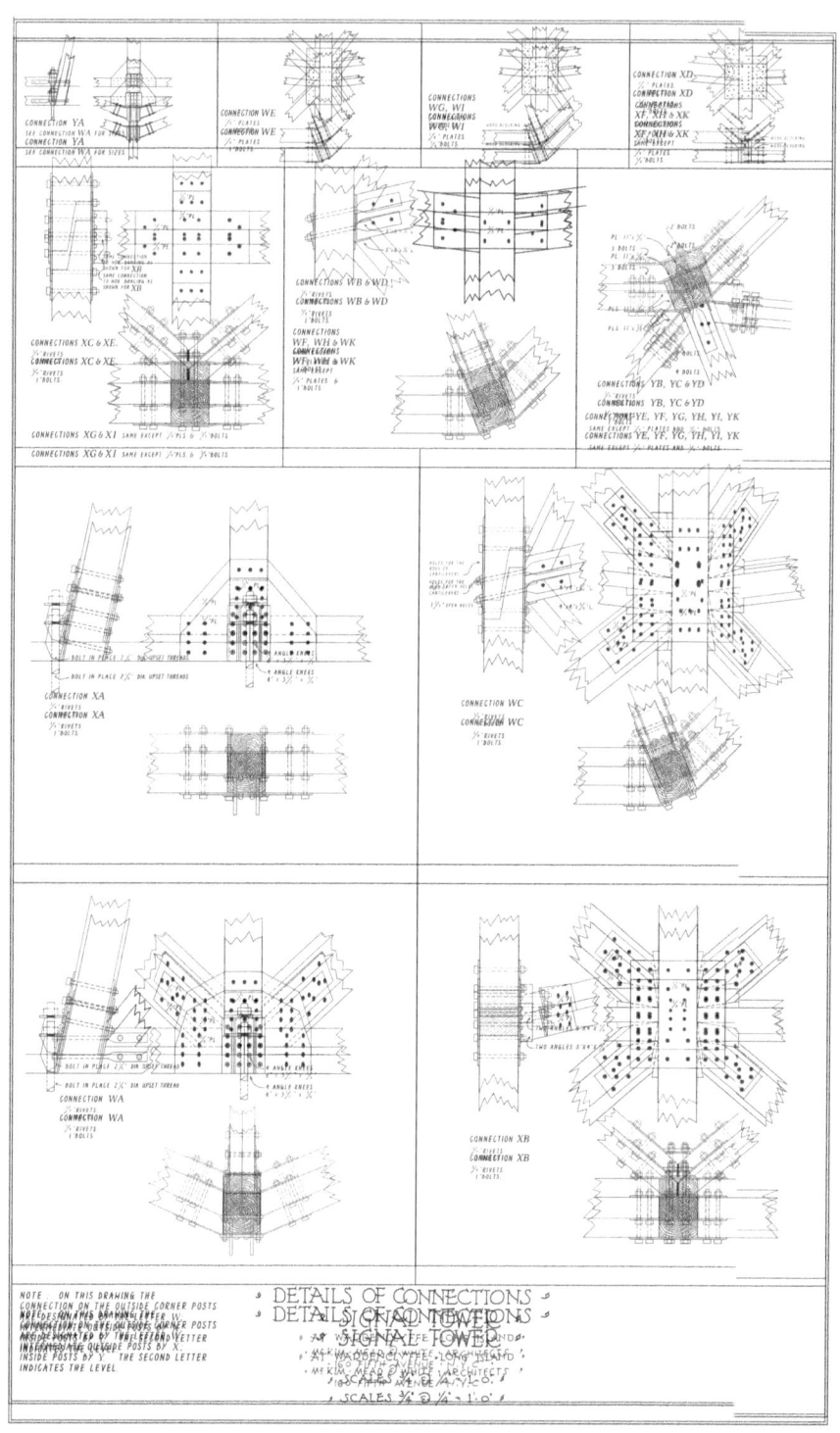

DETAILS OF CONNECTIONS
SIGNAL TOWER

SCALES ¾ & ¼ = 1'·0'

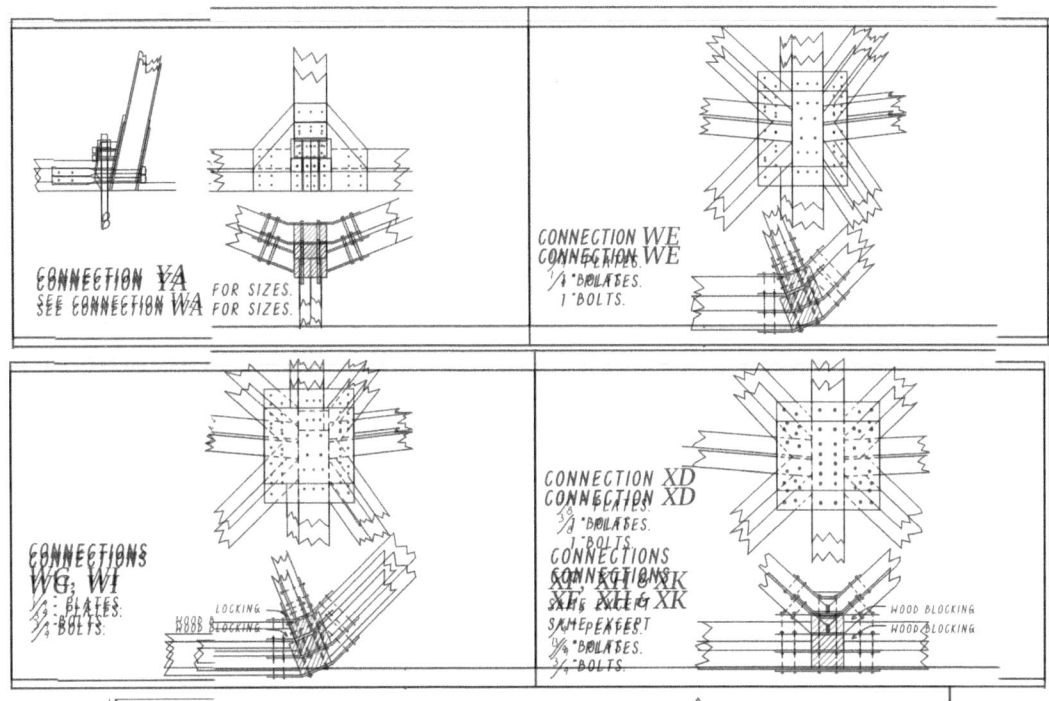

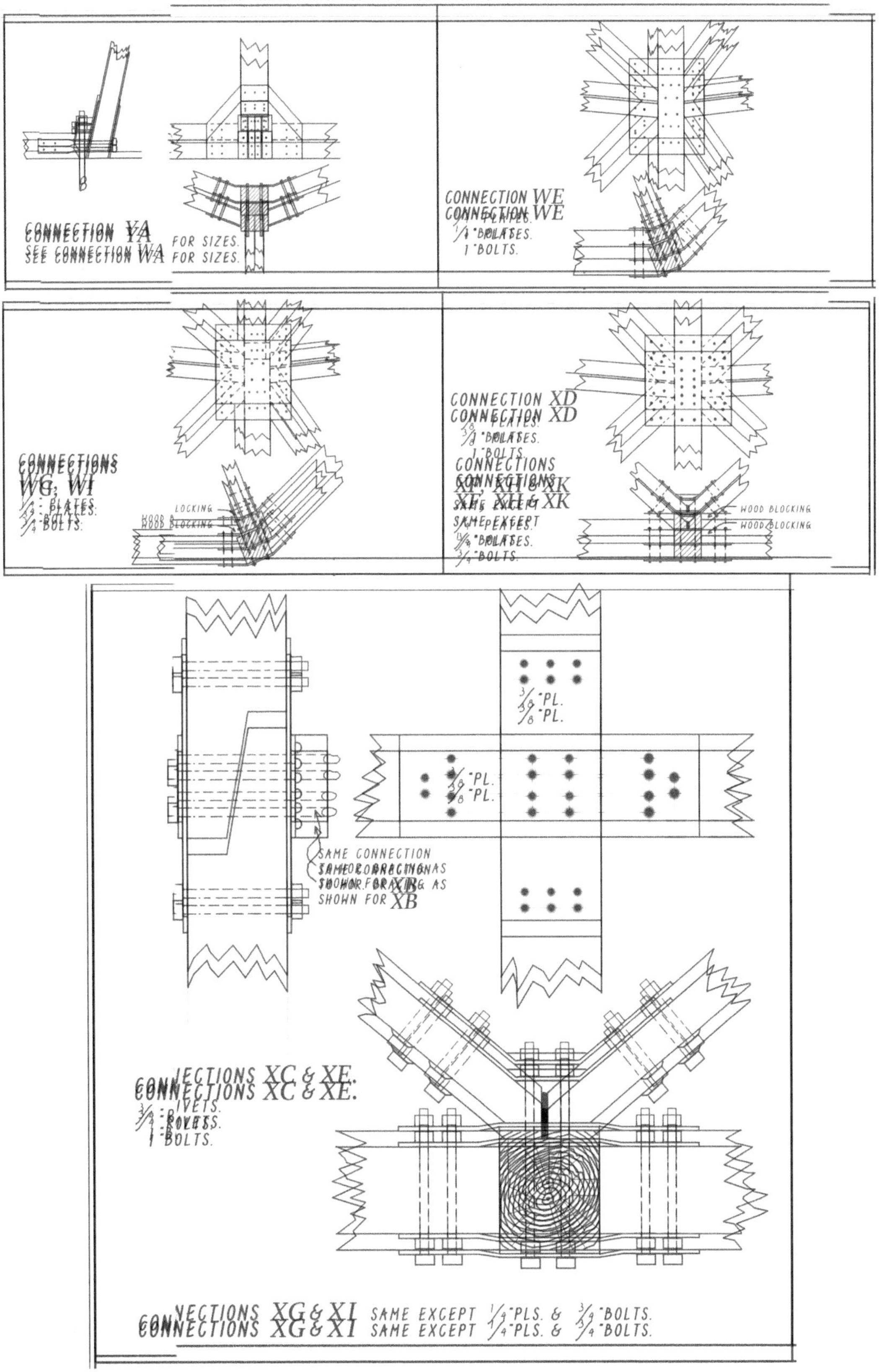

CONNECTION YA
SEE CONNECTION WA FOR SIZES.
CONNECTION YA
SEE CONNECTION WA FOR SIZES.

CONNECTION WE
CONNECTION WE
1/4 "PLATES.
1 "BOLTS.

CONNECTIONS WG; WI
1/4 "PLATES.
3/4 "BOLTS.

LOCKING
WOOD BLOCKING

CONNECTION XD
CONNECTION XD
3/8 "PLATES.
1 "BOLTS.

CONNECTIONS XF, XH & XK
CONNECTIONS XF, XH & XK
SAME EXCEPT
3/8 "PLATES.
3/4 "BOLTS.

WOOD BLOCKING
WOOD BLOCKING

3/8 "PL.
3/8 "PL.

3/8 "PL.
3/8 "PL.

SAME CONNECTION & BRACING AS SHOWN FOR XB
SAME CONNECTION & BRACING AS SHOWN FOR XB

CONNECTIONS XC & XE.
CONNECTIONS XC & XE.
3/4 "RIVETS.
1 "BOLTS.

CONNECTIONS XG & XI SAME EXCEPT 1/4 "PLS. & 3/4 "BOLTS.
CONNECTIONS XG & XI SAME EXCEPT 1/4 "PLS. & 3/4 "BOLTS.

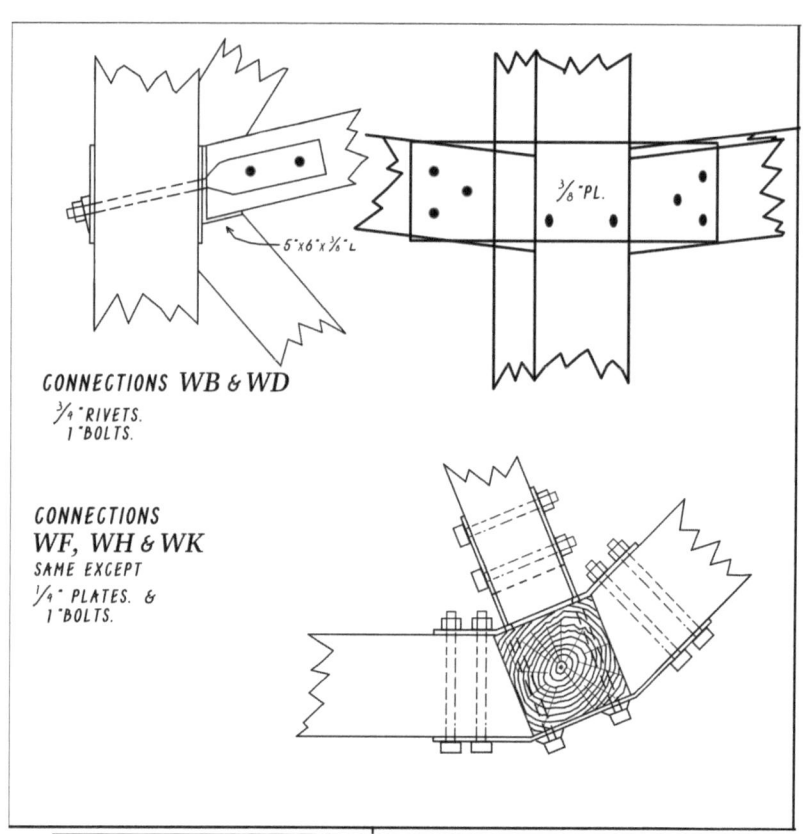

CONNECTIONS WB & WD
¾"RIVETS.
1"BOLTS.

**CONNECTIONS
WF, WH & WK**
SAME EXCEPT
¼" PLATES. &
1"BOLTS.

⅜"PL.

5"x6"x⅜"L

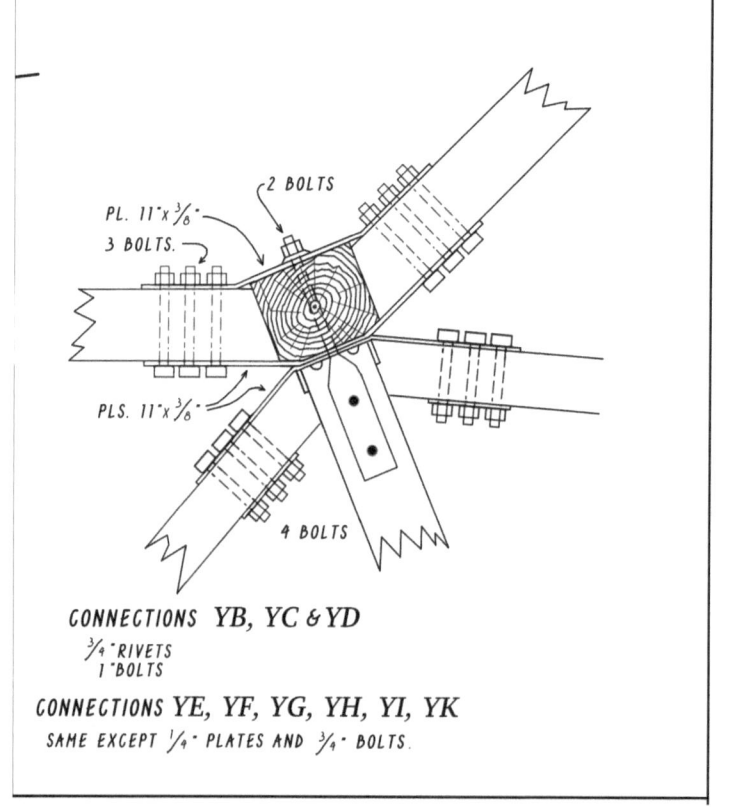

PL. 11"x⅜"
3 BOLTS.
2 BOLTS
PLS. 11"x⅜"
4 BOLTS

CONNECTIONS YB, YC & YD
¾"RIVETS
1"BOLTS

CONNECTIONS YE, YF, YG, YH, YI, YK
SAME EXCEPT ¼" PLATES AND ¾" BOLTS.

Tower Cupola - Old

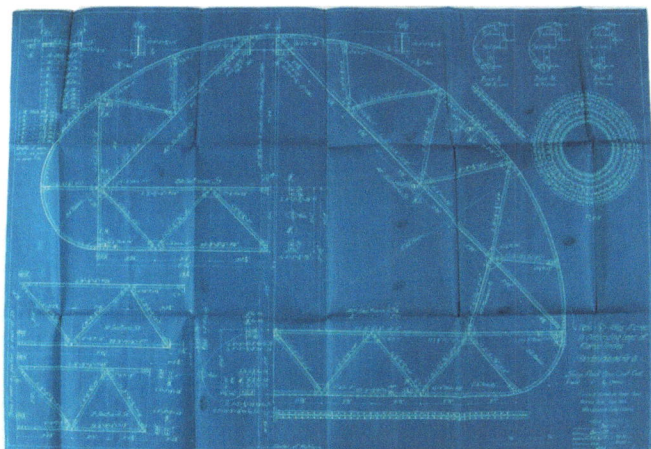

This blueprint was drawn at a scale of 1 inch for 1 foot, but this is not stated on the blueprint.

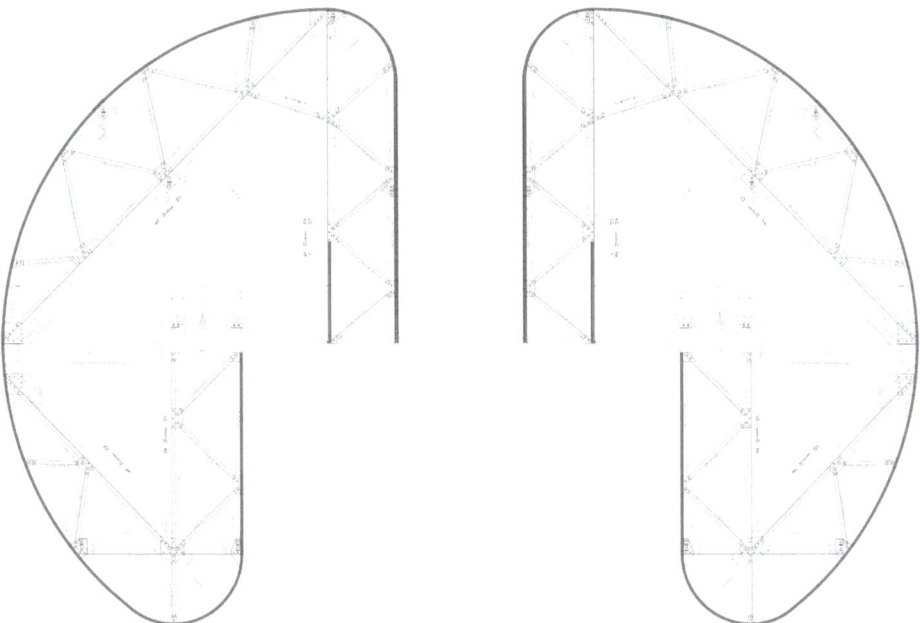

From this blueprint, we can see the original cupola design was much smaller. You can also clearly see the cylindrical hole in the middle from where the "safety valve" could rise.

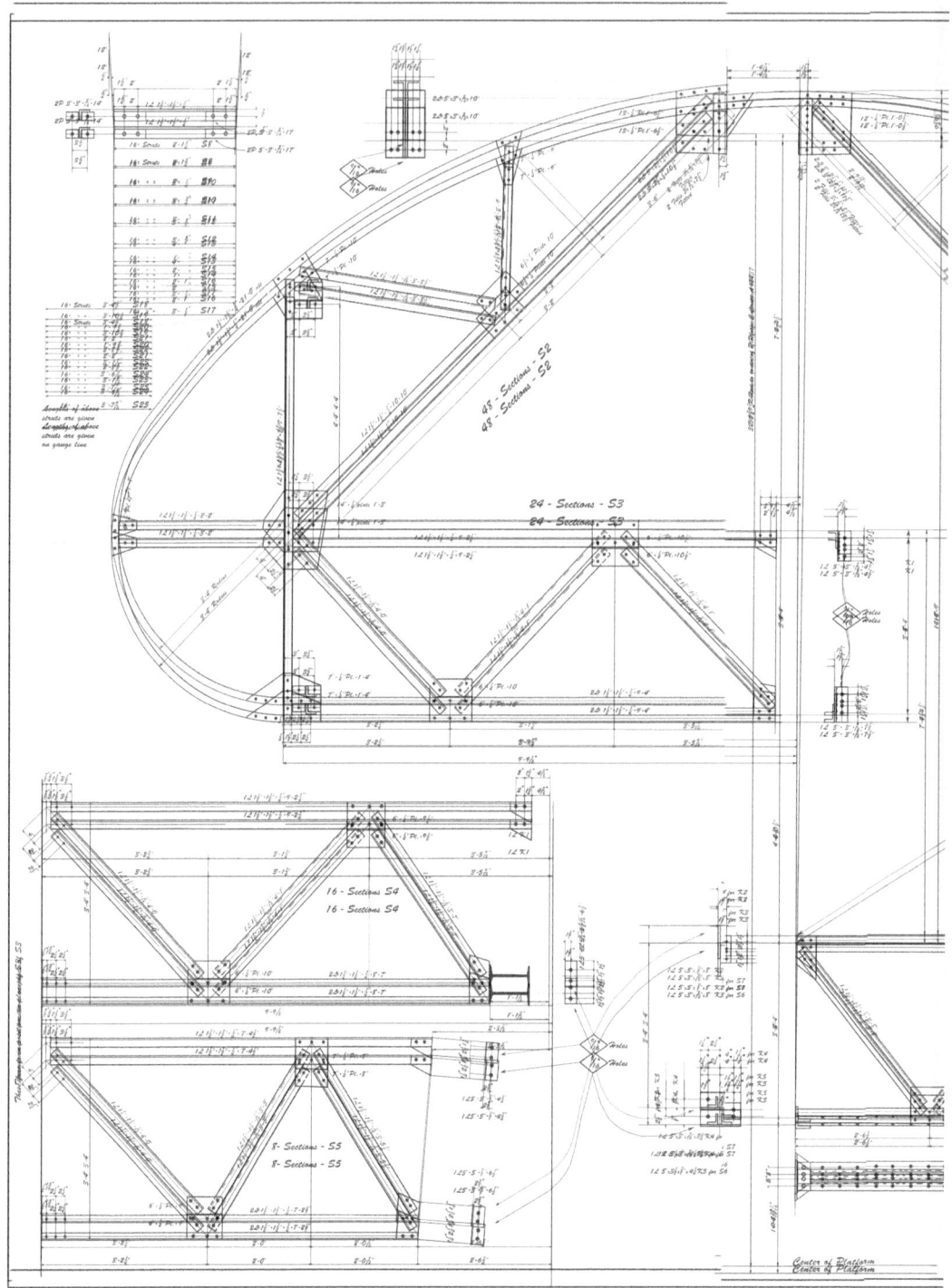

The yellow line on the right is the overlap with the image on the next page. There is no such line in the original.
The yellow line on the right is the overlap with the image on the next page. There is no such line in the original.

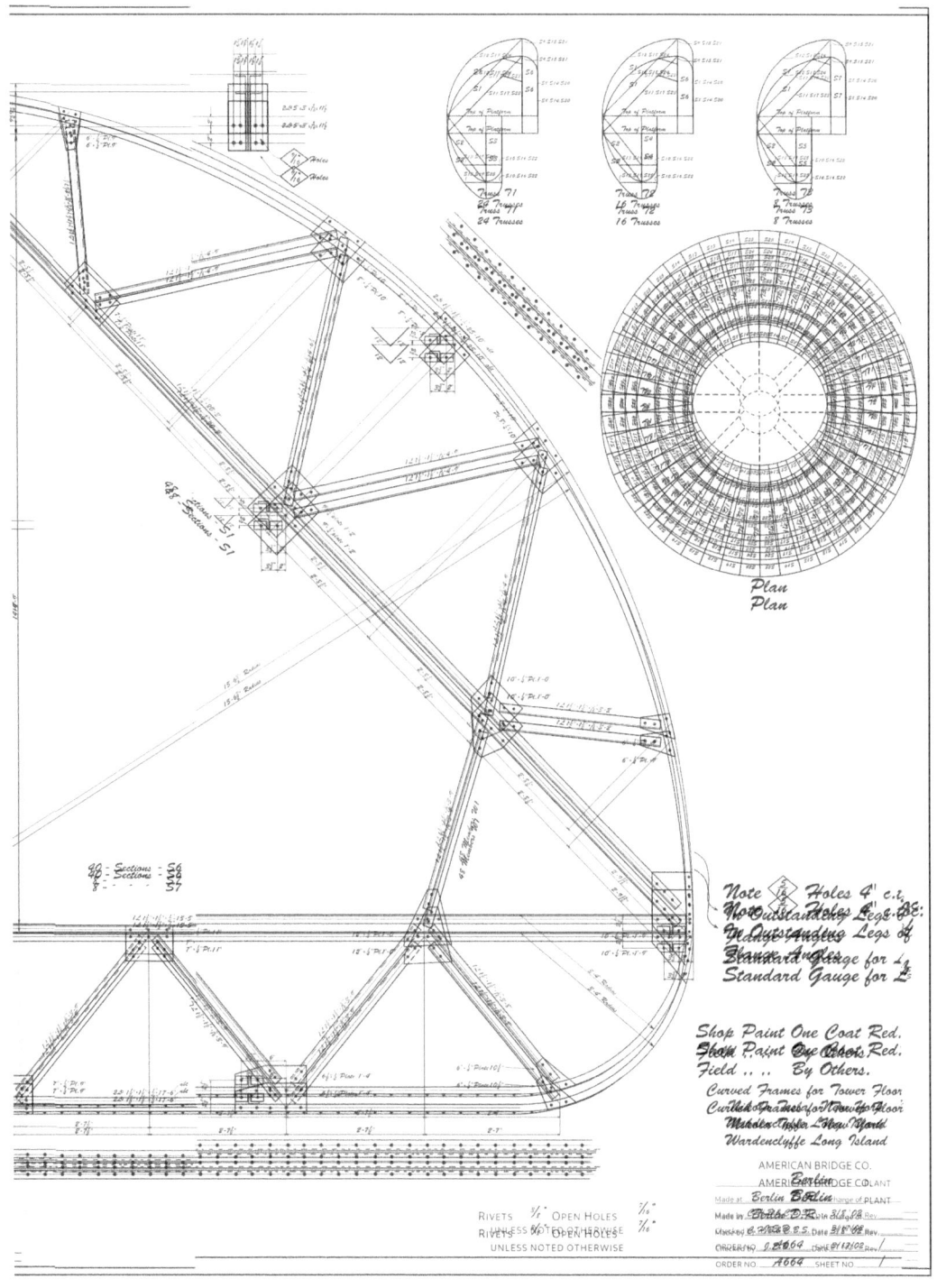

Plan
Plan

Note ⬦ Holes 4' c.t.
In Outstanding Legs of
Standard Gauge for L_4
Standard Gauge for L_2

Shop Paint One Coat Red.
Shop Paint One Coat Red.
Field . . . . By Others.
Curved Frames for Tower Floor
Curved Frames for Tower Floor
Nikola Tesla New York
Wardenclyffe Long Island

AMERICAN BRIDGE CO.
AMERICAN BRIDGE CO.
Made at Berlin Berlin
Made in Berlin Berlin
Checked Date 3/18/03
Order No. A664

RIVETS ⅞" OPEN HOLES ⅞₀"
RIVETS ⅞" OPEN HOLES ⅞₀"
UNLESS NOTED OTHERWISE

ORDER NO A664   SHEET NO

Tower Cupola - New

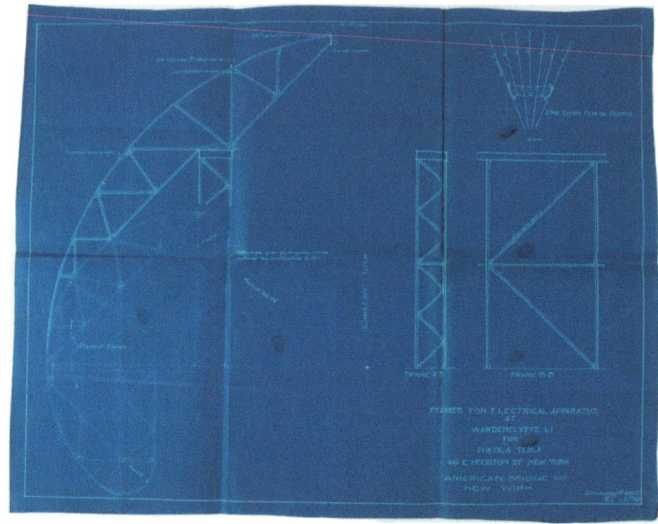

This blueprint was drawn at a scale of $^3/_8$ inch for 1 foot, but this is not stated on the blueprint.

The final design was almost twice as high and twice as wide.

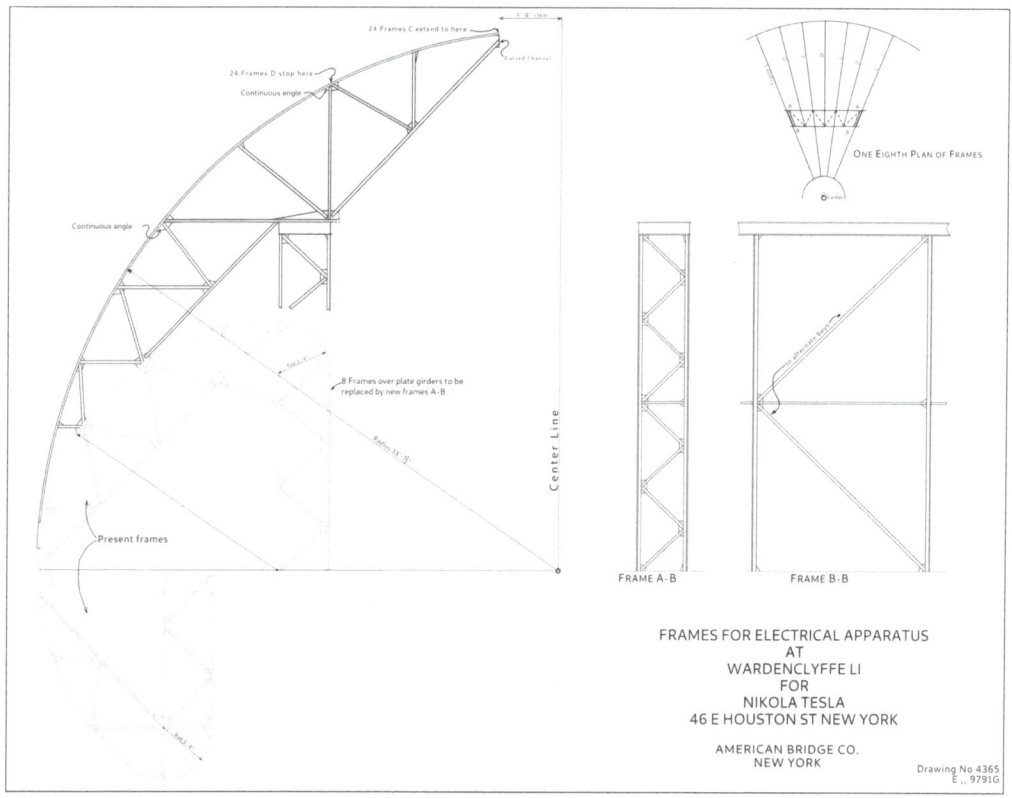

FRAMES FOR ELECTRICAL APPARATUS
AT
WARDENCLYFFE LI
FOR
NIKOLA TESLA
46 E HOUSTON ST NEW YORK

AMERICAN BRIDGE CO.
NEW YORK

Drawing No 4365
E., 9791G

This image is from the Electrical Experimenter of June, 1919, page 113.

On renderings like these, there is always this weird covering of the cupola. In the Colorado Springs Notes from July 24th, 1899 Tesla suggests using vacuum or low pressure hydrogen bulbs. Yet, from patent 1,119,732 and notes[3] from 1902, we

3   Document numbers 470.0189A and 470.0380A

learn that the entire cupola was to be covered with plates like this:

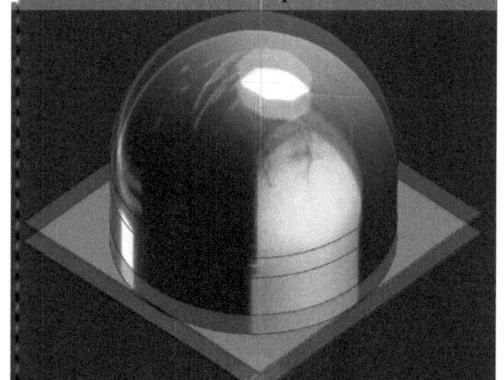

These are 9" x 9" plates with a 2" high cylinder in the middle, topped with half a sphere with a 4" radius. Tesla called these "shells". According to Tesla, this would significantly increase the capacitance of the cupola.

However, it would not at all look like those sketches, as 9x9" is so small on a cupola this size, that you would probably not even notice it.

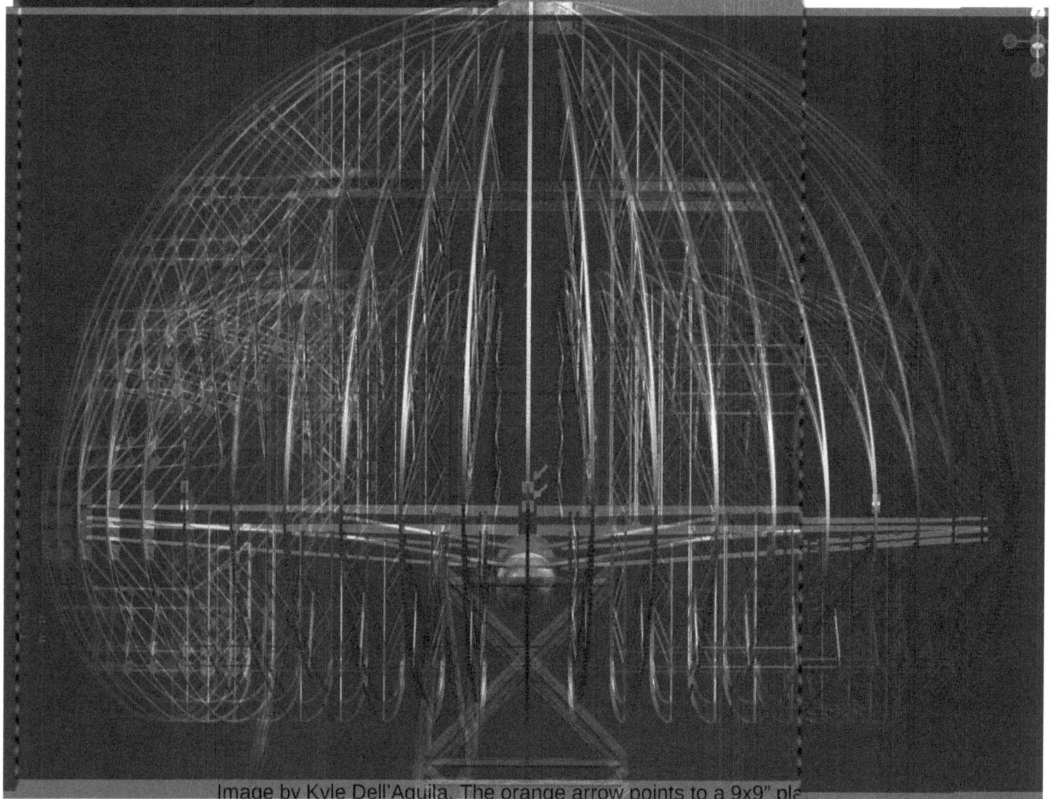

Image by Kyle Dell'Aquila. The orange arrow points to a 9x9" plate.

The sketch on the previous page is of a later date (1919). So, I think Tesla first (1899) wanted to use bulbs, then (1900) changed to plates, which are much more practical, and later again he may have changed back to bulbs. Something like this drawing to the left. But to match the sketch these bulbs would be 90-95 cm in diameter!

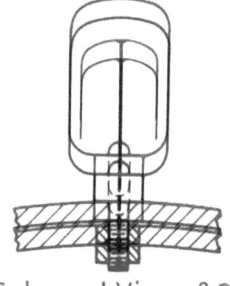

Enlarged View of One of the Attachments

This drawing can be found in document number 400.0064A and has been published in an article of May 16th, 1935, "The New Art of Projecting Concentrated Non-dispersive Energy Through Natural Media"

The cupola was never covered, so we cannot know what choice Tesla would have made in the end. We only know that he was playing with these two ideas. In the heat of our discussion on this subject, Marc checked with Kenneth L. Corum (who has studied Tesla's legacy for over 50 years). He more or less agreed with what I concluded above. I'll copy his full reply in the appendices of this book.

The cupola was never covered, so we cannot know what choice Tesla would have made in the end. We only know that he was playing with these two ideas. In the heat of our discussion on this subject, Marc checked with Kenneth L. Corum (who has studied Tesla's legacy for over 50 years). He more or less agreed with what I concluded above. I'll copy his full reply in the appendices of this book.

In this picture, we can see the alternative frame on the side away from the laboratory indicating that that is where the main entrance would be, conform the "artist impression" 2 pages back. And we can see stairs on the opposite side from where this picture was taken.

Tower in a more completed state.

Tower Floor Plan

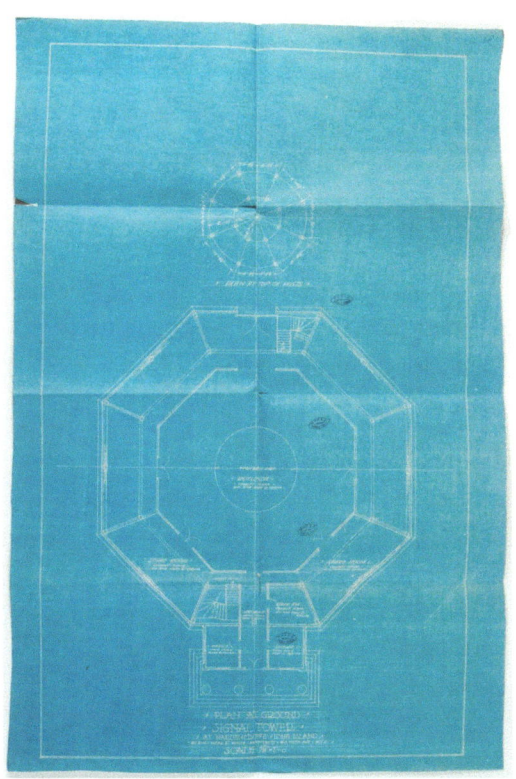

Besides the floor plan, the dimensions at the top of the posts are given here. Together with the foundation and the height of the posts, this is essential information if you want to model the tower. (or rebuild it?...)

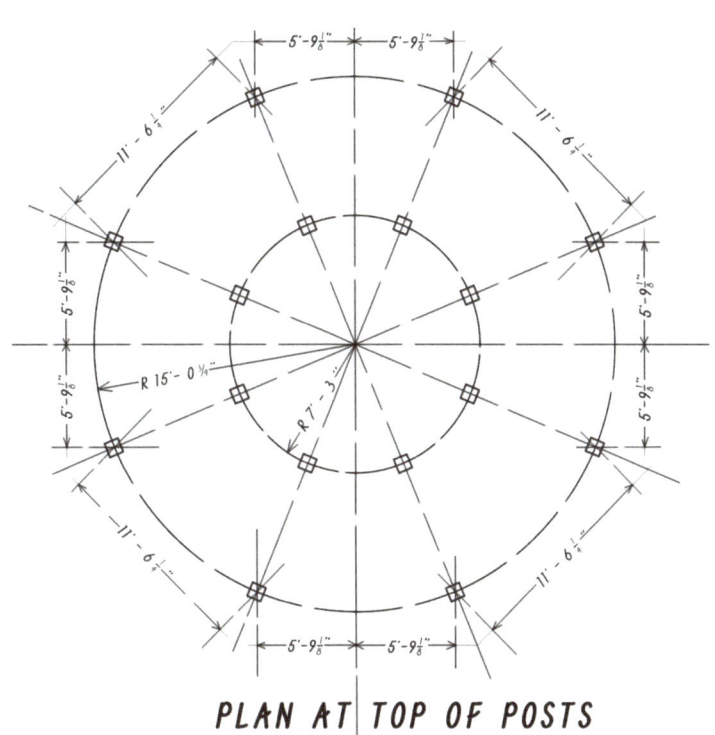

PLAN AT TOP OF POSTS

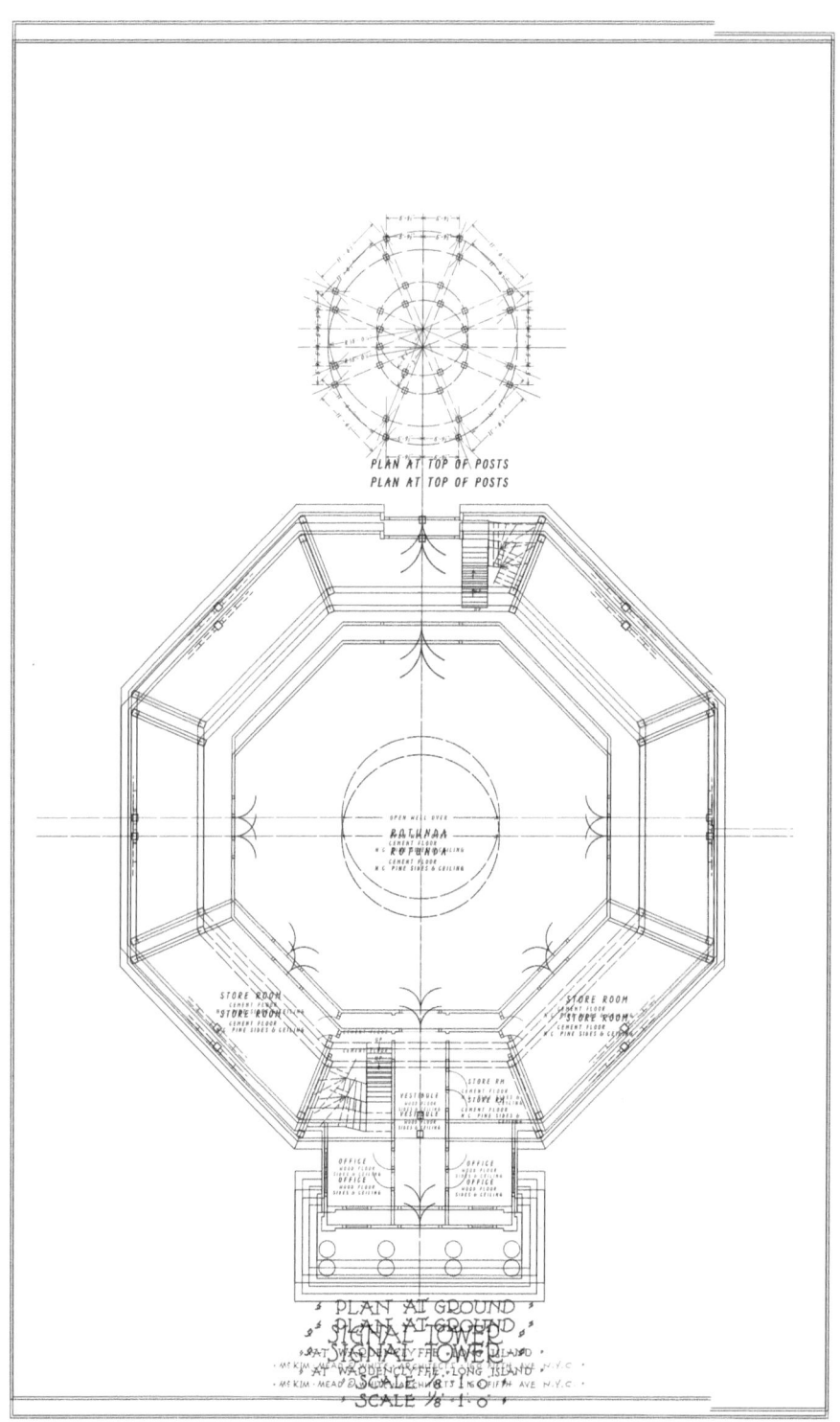

PLAN AT TOP OF POSTS
PLAN AT TOP OF POSTS

OPEN WELL OVER
RALHNAA
ROTUNDA

STORE ROOM
STORE ROOM

STORE ROOM
STORE ROOM

VESTIBULE
STORE RM

OFFICE
OFFICE

OFFICE
OFFICE

PLAN AT GROUND
PLAN AT GROUND
SIGNAL TOWER
SIGNAL TOWER
AT WOODCLYFFE LONG ISLAND
Mc KIM MEAD & WHITE 1 160 FIFTH AVE N.Y.C
SCALE 1/8 1 0

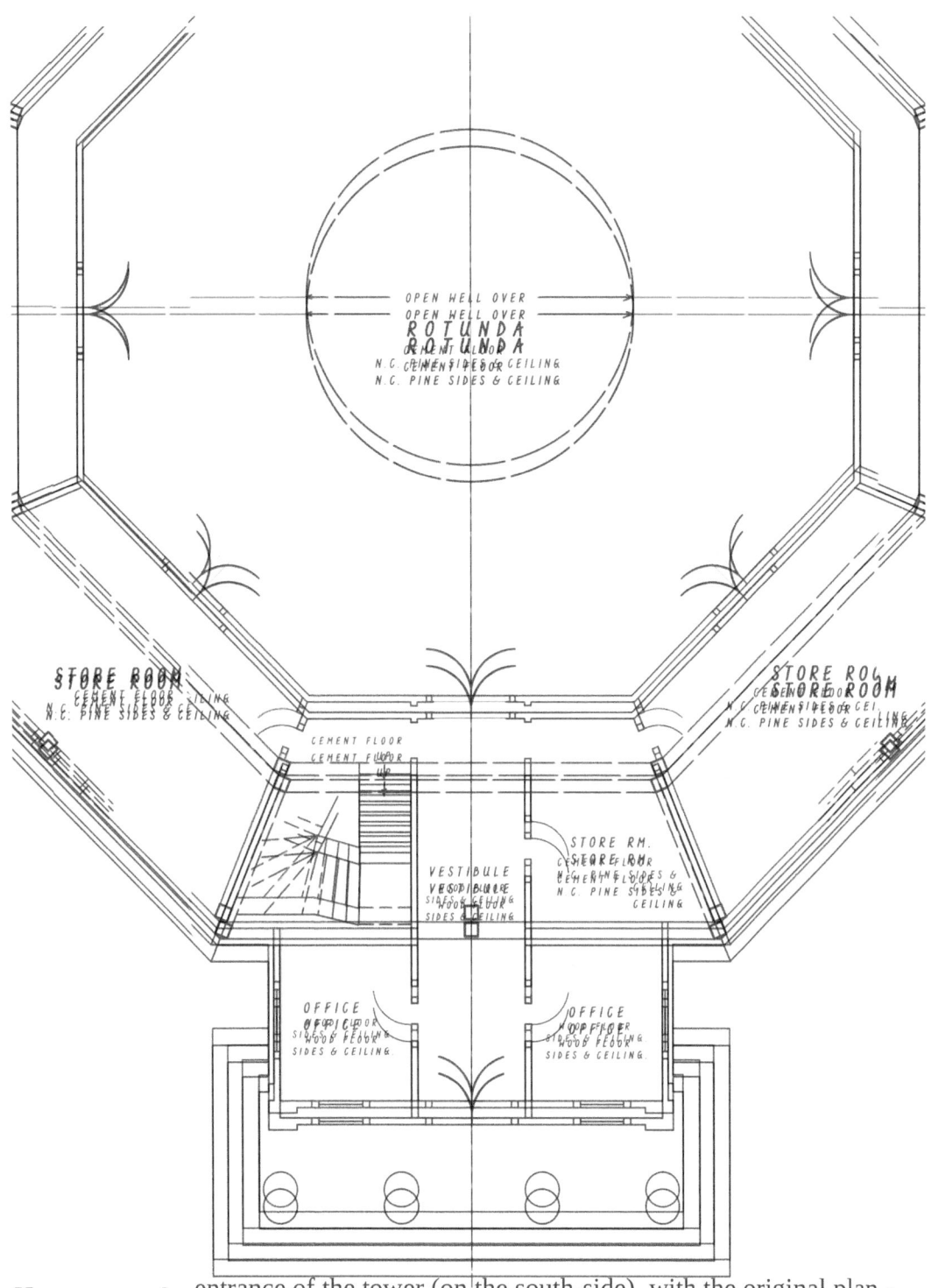

OPEN WELL OVER
OPEN WELL OVER
ROTUNDA
ROTUNDA
N.C. CEMENT FLOOR CEILING
N.C. PINE SIDES & CEILING

STORE ROOM
CEMENT FLOOR CEILING
N.C. PINE SIDES & CEILING

STORE ROOM
STORE ROOM
CEMENT FLOOR
N.C. CEMENT FLOOR CEI
N.C. PINE SIDES & CEILING

CEMENT FLOOR
CEMENT FLOOR
up

STORE RM.
CEMENT FLOOR
CEMENT FLOOR
N.C. PINE SIDES &
N.C. PINE SIDES &
CEILING

VESTIBULE
VESTIBULE
WOOD FLOOR
WOOD FLOOR
SIDES & CEILING
SIDES & CEILING

OFFICE
OFFICE
WOOD FLOOR
WOOD FLOORING
SIDES & CEILING.

OFFICE
OFFICE
WOOD FLOOR
WOOD FLOORING
SIDES & CEILING.

Here we see the entrance of the tower (on the south-side), with the original plan for the stairs:

Laboratory Cross Section

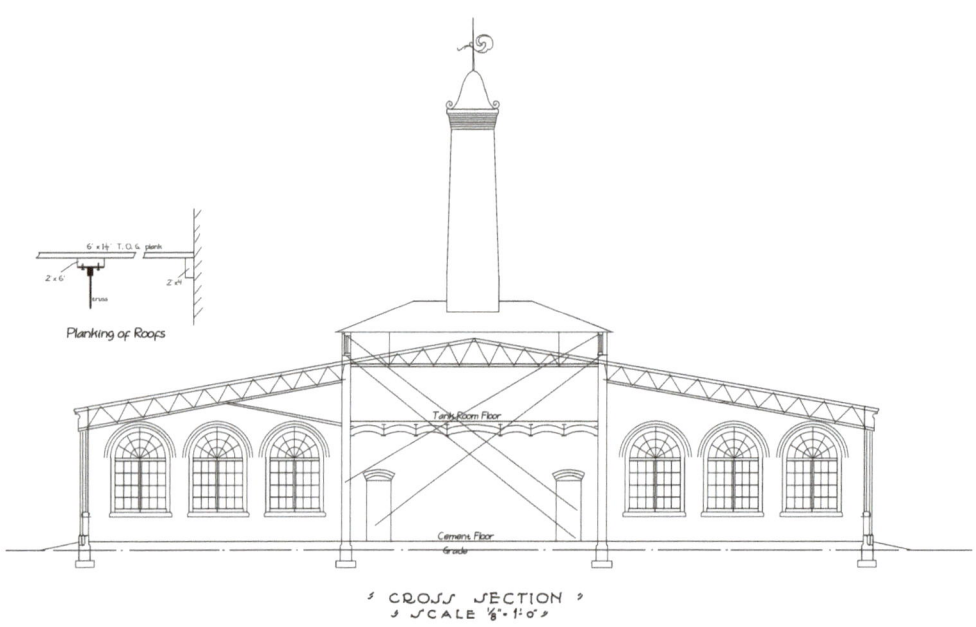

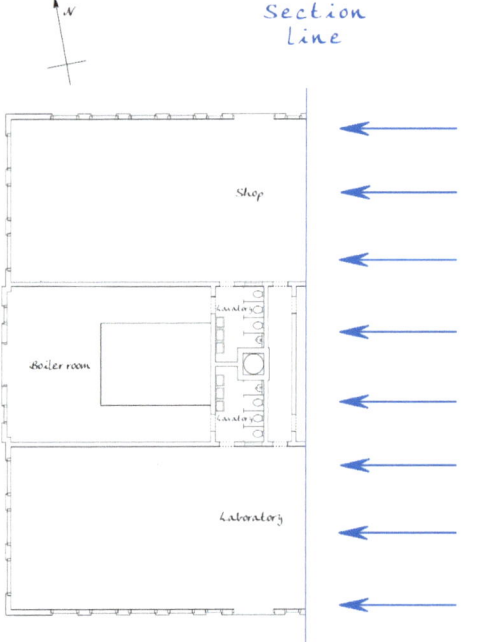

It appears that this cross section has been made on this line.

You can see the two doors to the hallway connecting the shop to the laboratory.

Laboratory Front Elevation

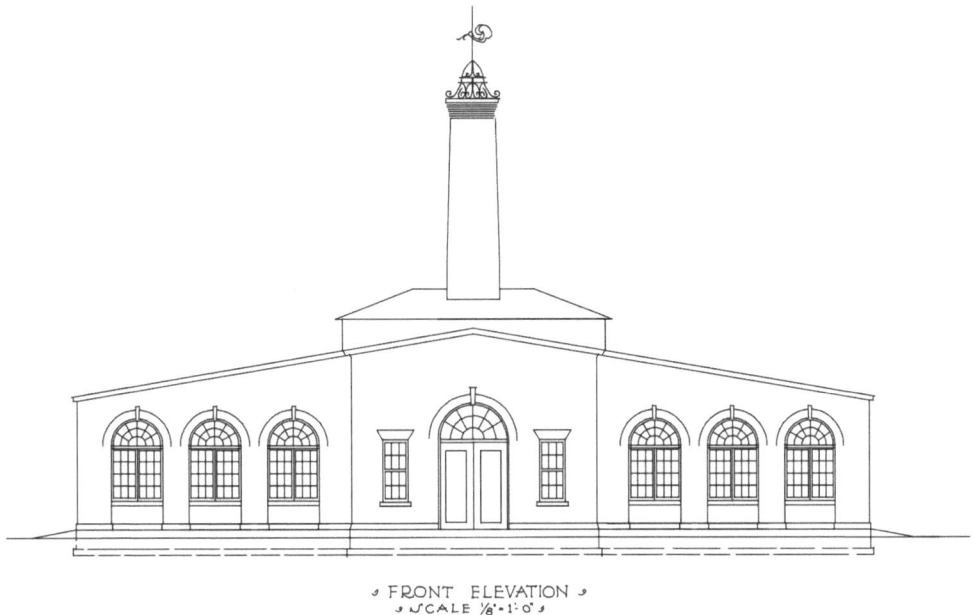

⸴ FRONT ELEVATION ⸴
⸴ SCALE ⅛"=1'0" ⸴

This one is called "Front Elevation" but in reality, it is the view from the side, as shown in this photo taken from the West side.

Laboratory Floor Plan

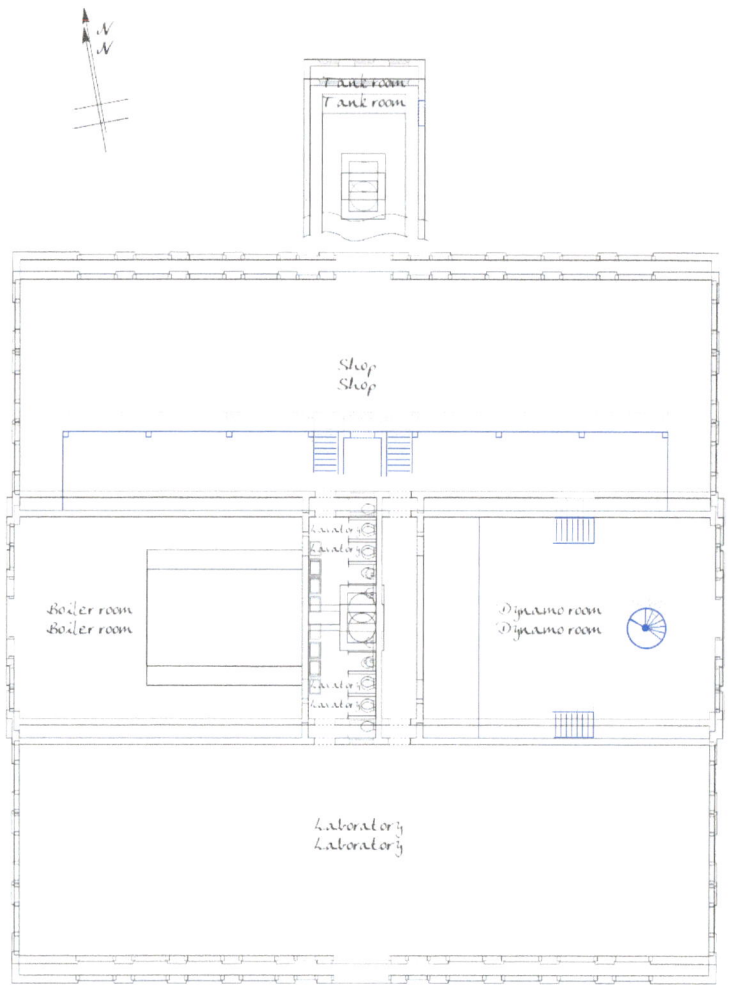

This appears to be an early plan. I have added, in blue, what can be seen in the pictures. Black is copied from the original document.

The boiler room is on the West side in this plan. But that does not make much sense, as the train passes on the East side. The final plan looks like a mirrored version of this. The dynamo room and boiler room switched places, and the lavatories moved to the East side.

This is corroborated by pictures of the laboratory that show an experimental set-up in the West side of the laboratory. As power is generated in the Dynamo room it makes sense to do these experiments near this room.

There also is a 3D scan[4] of the building online showing the situation at the time the property was purchased by "the Tesla Science Center at Wardenclyffe".

I walked through that scan, carefully creating a map of the current situation. Below you can see the result. Because all of Tesla's equipment had been removed it is impossible to tell with certainty where the boiler and where the generator once stood. But you can see that the lavatories were on the East side, not the West as

4   Check QR codes in the back of this book.

depicted above:

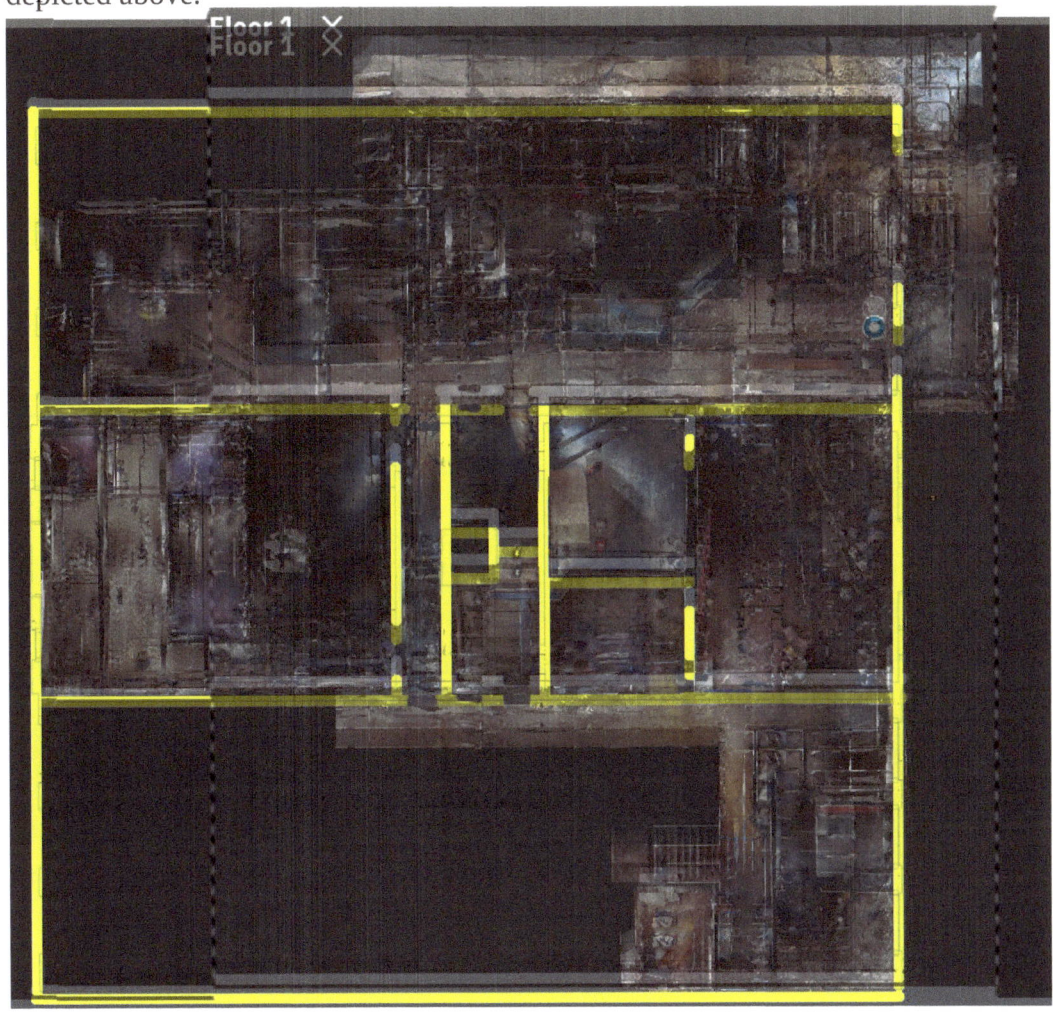

Floor 1 ✕

For these reasons, I think we can now say with certainty that the actual floor plan would have looked like this below.

There are only two things that I was unable to solve. The first is that there are 6 toilets here and 2 urinals, while in the "Wardenclyffe Foreclosure Appeal Proceedings" it is stated that there were 7 toilets and 1 urinal. So one of the urinals was replaced with a toilet, but I don't know which one. The other thing is that I don't know how the two doors to the shop, one from the lavatories and one from the hallway, were merged into one as can be seen in the pictures.

For these reasons, I think we can now say with certainty that the actual floor plan would have looked like this below.

There are only two things that I was unable to solve. The first is that there are 6 toilets here and 2 urinals, while in the "Wardenclyffe Foreclosure Appeal Proceedings" it is stated that there were 7 toilets and 1 urinal. So one of the urinals was replaced with a toilet, but I don't know which one. The other thing is that I don't know how the two doors to the shop, one from the lavatories and one from the hallway, were merged into one as can be seen in the pictures.

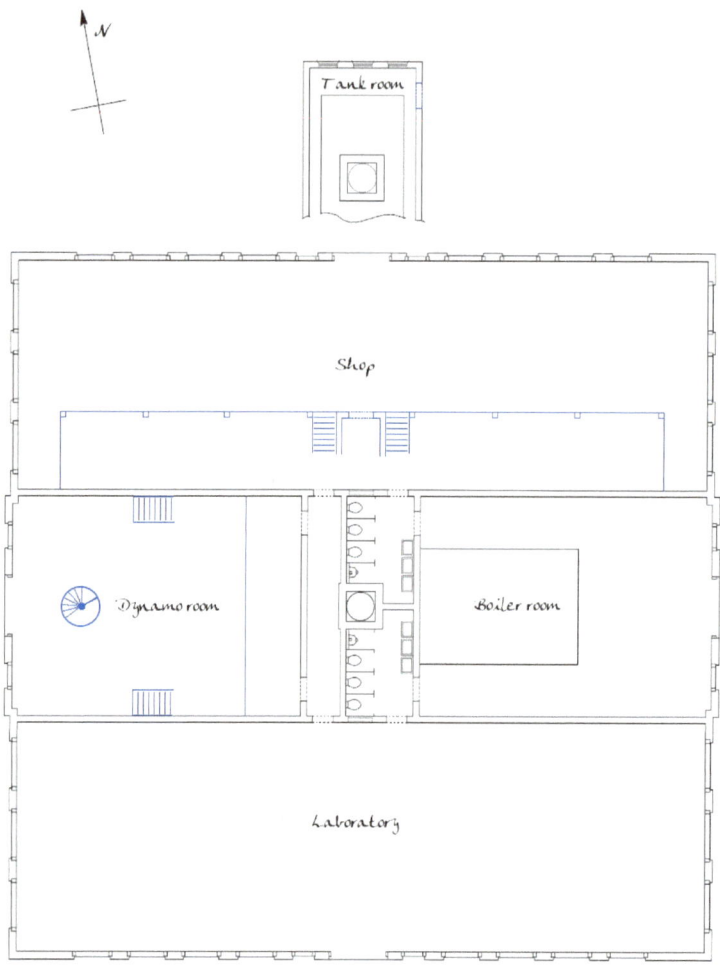

Here you can see the single door entering into the Shop. On either side of this door there is a staircase going up to the balcony.

Inside the Tower

From all the notes Tesla made during the construction of the tower, it appears that there still were many unanswered questions about the exact implementation of Tesla's plan, but there are a few things that we know for certain. So let's start there.

In 1997, Leland Anderson published a few diagrams in issue 26 of the "Electric Spacecraft" magazine in an article called "Rare Notes From Tesla on Wardenclyffe". These were diagrams that Tesla drew during this time and of which I have found the originals in the Tesla Museum.

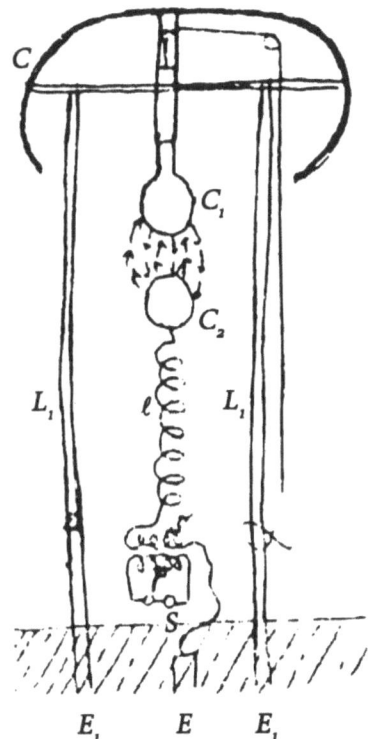

The article starts with a copy of a note (470.0042A) by Tesla, with a diagram and a lot of explanatory text, dated May 29th, 1901. This diagram is repeated in Fig. 4 in this article, and you can see it here to the left.

Figures 5 to 7 (see below) show similar diagrams but with a different driving circuit. Yet, in all cases we see the Cupola (C) being connected to ground (through L_1), with a movable sphere (C_1) below it onto which the driving circuit discharges.

These other diagrams are from documents 514.0016A, 514.0017A and 514.0020A. It is important to know that the originals do not contain any text and that the texts found in the "Rare Notes" were added by the Corum brothers. I have included "thumbnail copies" on the next page.

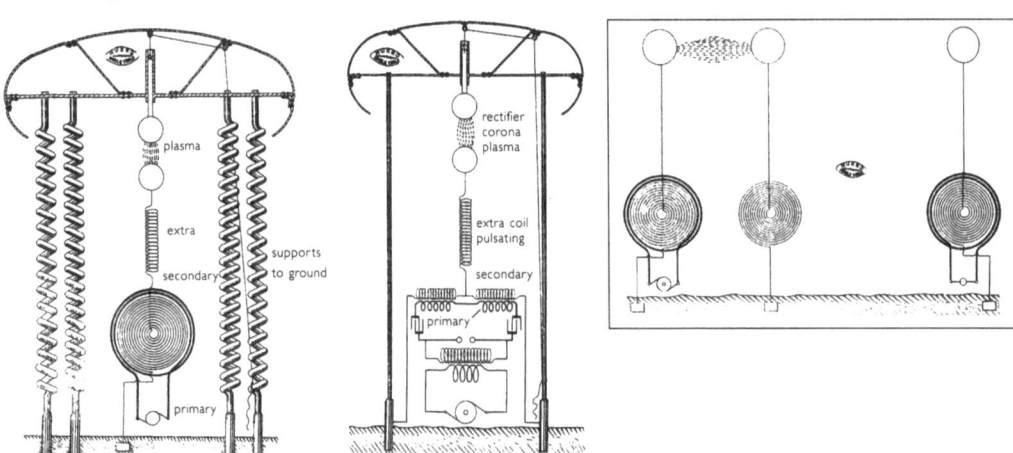

Each of these diagrams shows a movable sphere connected to the cupola, and in one of Tesla's notes (470.0137A) we read:

Jan. 12, 1906.
Curvature to be given to
upper movable small sphere
so that the streamers appear
at 10^7 Volts=p
q=Sxσ   S=4πr²   σ=20
q= 4πr² x 20
Capacity c = r

$$\frac{q}{c}=p=\frac{4\pi r^2\times 20}{r}=\frac{10^7}{300}$$

$$r=\frac{10^7}{300\times 20\times 4\pi}=\frac{10^4}{6\times 4\pi}$$

$$=\frac{10^4}{75.4}=\frac{10^5}{754}=132.63\,c.m.$$

$$=\frac{132.63}{30.48}=4.35\,feet$$

Opening is just a trifle too
small but probably right
on account of influence
of sphere: O.K.

From this, we now know that C_1 in the first diagram was to be a sphere with a 132 cm radius, and I think C_2 would be the same size.

In the middle of the tower, there was a shaft (10' x 12') going down 120 ft.[5] This shaft *was first covered with timber and the inside with steel and in the centre of this, winding stairs were going down and in the centre of the stairs there was a big shaft again through which the current was to pass.*[6] These steel plates were part of the grounding system for the tower. From Tesla's notes, it looks as if Tesla was planning to use a half wave resonator ('extra coil') with one end capacitively coupled to ground and the other end connected to C_2 which was to discharge onto C_1 when it reached 10 million Volts.

This could also solve the issue of extremely high potentials inside the tower as I will demonstrate in the last chapter.

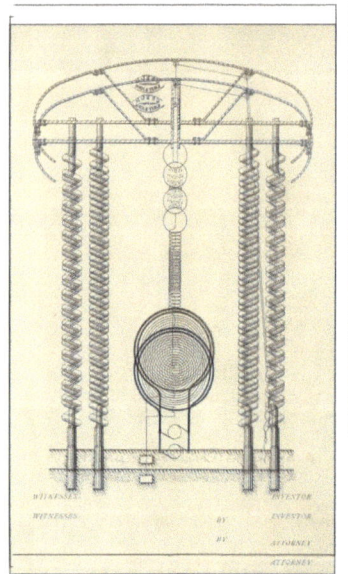

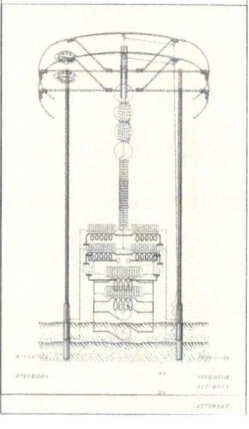

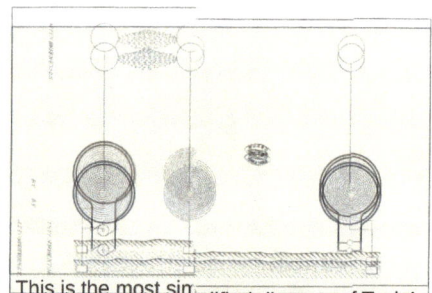

This is the most simplified diagram of Tesla's Magnifying Transmitter. Note that all three drawings are made on paper intended for patent applications, but Tesla never patented this:

_____

[5] 10 ft = 3.048 m, 12 ft = 3.6576 m, 120 ft = 36.576 m
[6] Quoted from Foreclosure Appeal (line 531).

The Grippers

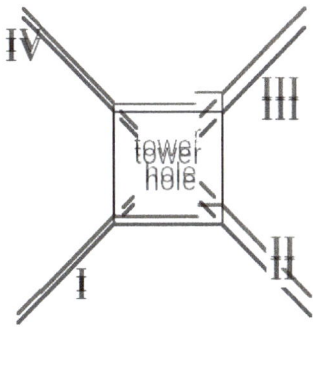

A short note on the grounding system. From notes[7] dated April 26, 1903, we can see there were four 2.5" pipes pushed from the bottom of the shaft in 4 directions, roughly NE, SE, SW and NW.

In the Foreclosure Appeal, we read:

I had special machines rigged up which would push the iron pipe, one length after another, and I pushed these iron pipes, I think sixteen of them, three hundred feet, and then the current through these pipes takes hold of the earth.

From the notes, we learn that these 4 pipes are about 152 feet, half of what is mentioned above. It is possible that Tesla thought of these pipes in pairs, NE+SW and NW+SE, of 300 feet each and connected in the middle; or he may have doubled their lengths later.

If 16 pipes are 300 ft (91.44 m) then 1 pipe is 18¾ feet (5.715 m). This looks plausible to me.

In recent years this place has been scanned with ground penetrating radar and in the results you can easily identify these 4 pipes, but it looks like there are many shorter pipes between the NE and SE pipes. Yet, I don't know to what extent we can trust these results. Until someone goes down there into the tunnels and proves otherwise, my bet is on just these 4 pipes, each being about 150 ft. long. These pipes were electrically connected to the iron plates that covered the inside of the 120 ft. well below the tower.

_____

7 Doc. nr. 250.0005A and 250.0006A show the location, 250.0008A mentions the size.

Inside the Laboratory

The information below is compiled directly from Tesla's words (in blue italics) during the Foreclosure Appeal in 1922.

Dynamo Room

1 400 HP Westinghouse auto compound (reciprocating engine) No. 1497, size 16 by 27 by 16

1 direct connecting Westinghouse alternating current generator 200 KW., Serial No. 155407, complete with lubricator, gauge, Rheostadt, switchboard and switches

1 35 KW Westinghouse engine, No. 4750, size 8½ by 8, with direct connected double current generator, 25 KW., Serial No. 168362, complete with lubricator, gauge, Rheostats, switchboard and switches

1 15 H. P. Westinghouse motor, No. 162315

4 Westinghouse transformers, 15 kw. type O. D.

1 tank manufactured by Stoutenborough

1 truck

1 Fairbank's scale

1 Laidlaw Dunn-Gordon pump, No. 16473

The first 2 items are to provide power to the tower, the 3rd provides power to this building.

There was a high pressure compressor which also formed an essential part of the equipment. And then there was a low pressure compressor or blower. Then there was a high pressure pump and a reciprocating low pressure pump. And of course this compartment also contained the switches and the switchboard and all that which goes with the equipment of the plant. Then there was a gallery on the top on which certain parts were placed and arranged that were needed daily in the operation.

Workshop

I have only been able to find one picture of the workshop, which is unfortunate as a lot has been added since the blueprints were drawn.

1	Westinghouse electric motor, used for power to drive machine shop, type C, induction motor, 6 H. P., No. 162319
1	Milling machine with tools complete, made by Brown & Sharp Manufacturing Company
1	lathe made by Pond Machine Tool Company, No. P-3040, with tools, belting and shafting
11	work benches
4	vises
1	Westinghouse, type C, 2 H. P. induction motor, No. 162278
1	Westinghouse, type C, induction motor, 2 H. P. Serial No. 162272
1	Westinghouse, type C, induction motor, 5 H. P., No. L-74487
1	Westinghouse motor, about ¼ H. P., No. 22190
3	lathes made by F. E. Reed of Worcester, Mass. with shafting, belting and tools
1	plainer made by Hendey Machine Co., with shafting, belting and tools
1	plainer made by Pedrick & Ayr, with shafting, belting and tools
1	F. E. Reed, hand drill press, shafting, belting and tools
1	large drill press by Prentice Brothers, with shafting, belting and tools
36	lockers containing miscellaneous supply of valves, joints, lubricators, fittings, scales, switches, single and double pole, socket, wrenches, fuses and plugs
1	testing fan motor
	A quantity of telephone and bell wire
	A quantity of lead cable material
4	radiators
	A quantity of drills, rose bits, reamers, taps, and all tools for milling machine and lathes, at present time in store room located in said workshop
2	oil tanks
1	testing motion by Crocker Wheeler, ½H.P. with Rheostat, No. 1000
1	submarine boat
1	clock

This compartment contained, I think, eight lathes ranging in swing from eight inches to thirty-two, I believe. Then there was a milling machine and there was a planer, and shaper, a spliner, a vertical machine for splining. Then there were three drills, one very large, another medium and a third quite small one. Then there were four motors which operated the machinery. Also a grinder and an ordinary grindstone, a blacksmith's forge. Then a special high temperature stove and the blower for the forge.

Boiler Room

2 300 HP Babcock & Wilcox boilers with steam gauges and water columns and with Metropolitan injector and Worthington feed pump
1 other feed pump
1 hand blacksmith and forge

Lavatories

7 toilets
1 urinal
6 wash basins

Tank Room

On a sort of second floor in the middle of the building *there were big water tanks that were placed around the chimney so as to utilize some of the waste heat. These tanks (made of quarter inch thick, galvanized, steel) had a capacity of about 16,000 gallons[8].* This room could be accessed from the balcony in the dynamo room.

Almost 73,000 litres.

Laboratory

7   Rheostats
4   desks
2   safes
3   motors
1   set of storage batteries and tanks
1   submarine boat
1   Westinghouse motor, No. 28292
1   Westinghouse motor, type C, 5 H.P. No. 62320
1   Westinghouse motor, type C, 5 H.P. No. 22070
4   high-tension transformers in tanks; and switchboards
    Wiring drums
    Drafting boards and tools
24 chairs
2   clocks
14 radiators

Right along the back wall that separated this compartment from the rest of the building there were two special glass cases in which I kept the historical apparatus which was exhibited and described in my lectures and scientific articles. There were probably at least a thousand bulbs and tubes each of which represented a certain phase of scientific development. Then close, beginning with these two glass cases, there were five large tanks. Four of those contained special transformers according to my design, made by the Westinghouse Electric Manufacturing Company. These were to transform the energy for the plant. They were about, I should say, seven feet high and about five by five feet each, and were filled with special oil which we call transformer oil, to stand an electric tension of 60,000 volts. Then besides these four tanks there was another similar tank which was for special purposes, containing a transformer. Then there were two doors, one door that led to the other compartment and the other one led in the closets, and between those two doors there was a space on which was placed my electric generating apparatus. This apparatus I used in my laboratory demonstrations in two laboratories before, and I had also used it in the Colorado experiments where I erected a wireless plant in 1889[9]. That apparatus was precious because it could flash a message across the Atlantic, and yet it was built in 1894 or 1895. That is a complicated and very expensive apparatus.

Then beyond the door there were again four tanks, big tanks almost the same size as those I described. These four tanks were to contain the condensers, what we call electric condensers, which store the energy and then discharge and make it go around the world. These condensers, some of them, were in an advanced state of construction, two, I think, and the others were not. They were according to a

---

9   In the court documents it says 1889, which in combination with the next sentence is obviously impossible: He could not have used something in 1889 that he built in 1894 or 1895. The correct year is 1899.

principle of discovery. Then there was a very expensive piece of apparatus that the Westinghouse Company furnished me; only two of this kind of apparatus were made by the Westinghouse Company, one for me and one for themselves. It was developed together by myself and their engineers. That was a steel tank which contained a very elaborate assemblage of coils, an elaborate regulating apparatus, and it was intended to give every imaginable regulation that I wanted in my measurements and control of energy. Then on the last side, where I had described the first four big tanks there was a special 100-horse power motor and this motor was equipped with elaborate devices for rectifying the alternating currents and then sending them into the condensers. On this apparatus alone I spent thousands of dollars. The 100-horse power motor was specially constructed for me by the Westinghouse Company, but the other parts were all made by myself and that took a considerable portion of space there and it was a wonderful piece of apparatus.

I have photographs of these which will make this description very clear. Then along the center of the room, I had a very precious piece of apparatus. That was a boat which was illustrative of my discovery of teletaumatics; that is a boat which was controlled without wire, which would do anything you wanted, but there was no connection. This boat was exhibited by me on many occasions.

This picture shows a test set-up, 4 transformers in the back, 4 capacitors on the tables and a 100 HP synchronous rectifier in the middle. This test used the flat spiral coil that was hanging from the ceiling.

Then there were on each side long specially made, how do you call them, not desks or shelves, but closets, I might say, which were specially made to contain the apparatus, because I had accumulated for years hundreds of different kinds of appliances which stand for a certain principle, and this apparatus was stored in - there, and on top of these I had again all full of apparatus, each representing a different phase. And then on one side there were the desks and then on the other side there were the drawing implements and tools. And then in the corner, when you looked at the railroad side, on the right side in the corner there was my testing room and that contained - there were two precious instruments among these that Lord Kelvin made especially for me. He was a great friend of mine. A device for measurement invented by him; it is called a breach; and another a voltmeter of his. Both of these things were given to me and prepared for me by his special instructions. There were a lot of other instruments, voltmeters, wattmeters, ampere meters; in that small space there was a fortune in there.

This is an earlier picture that shows the desks and submarine in the back.

Conclusions

Let me first get note 470.0042A:
Let me first get note 470.0042A:

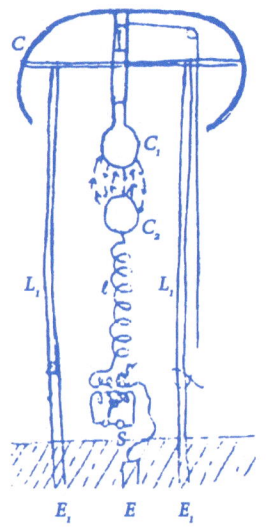

New York May 29, 1901
from old note

In annexed sketch a terminal C in form of a roof is supported on <u>conducting</u> supports L_1L_1. Terminal C_1 is adjustable and in contact with structure of roof or terminal C. A resonating system $C_2 \ell SE$ discharges into C_1 and produces oscillations in system $CL_1L_1E_1E_1$. This arrangement obviates necessity to support roof or terminal C on <u>insulated</u> supports.

Now in a sketch a scheme the difficulty will be probably to get the oscillations of the free system $CL_1L_1E_1E_1$ <u>slow</u> enough to be very effective in transmission through Earth as in my system. The length of conductors in the free system should be λ/4, and the length of the discharging current should be 3/4 λ or n/4 λ eventually, <u>n</u> being uneven number. Suppose, to get an idea, we take C = 10,000 cm. This is realizable. Then we have $\frac{2\pi}{10^3}\sqrt{L \times \frac{10,000}{9 \times 10^5}}$ the period of system. We should have vibration not much quicker than 100,000 and to satisfy this L would have to be: $\frac{1}{100,000} = \frac{2\pi}{10^3}\sqrt{\frac{L}{90}}$

L = 9 x $I0^5$/4 = 225,000 cm. Calculated it would appear that the supports L would have to be about 600 feet. The arrangement would be OK with quick oscillation. The self-induction of a straight conductor is L' = 2l'[$\log_e$ (2l'/r) - 0.75]. Now, take l'= 300 ft = 9000 cm. If we were to use iron pipes 4" diam. r = 5 cm. Then 2l'/r = 3600 and from this I find L' = <u>134,000</u> cm. Again taking the length 600 ft we would get inductance probably 268,000 cm. To get lower frequencies, evidently in above scheme self-induction must be increased.

P.S. The charging and discharge current may even be of different period and both vibrations used to excite receiver.

As is clear from the above, there were 2 resonating systems; $C_2 \ell SE$, the driving system and $CL_1L_1E_1E_1$ the free system. I think that the spark-gap between C_1 and C_2 was supposed to fire at a rate of 11.7 Hz or a multiple thereof. This would create the standing wave in the Earth's electricity that would facilitate the transmission of power and messages. The free system would be best for power transmission while the driving system would be best for messaging.

How were these to be implemented in Wardenclyffe? Let's start at the beginning.

In the middle we see the 200KW generator, directly behind it is the 400HP reciprocating engine

One of the 300 HP[10] Babcock & Wilcox boilers would produce pressurised steam to drive the 400 HP[11] Westinghouse auto compound, reciprocating engine. These two machines form a steam engine. Directly connected to the auto compound there was a 200 KW Westinghouse alternating current generator. This current would be used to power a Tesla coil, to produce a high frequency, high voltage output.

The exact diagram that Tesla had in mind for this, I was unable to find. But we know that there were 4 high voltage transformers and a synchronous rectifier that he could have used. There also were 4 capacitors in tanks on a kind of table, but those would not be the primary capacitors. These primary capacitors would be made of glass plates as we can read in several notes.

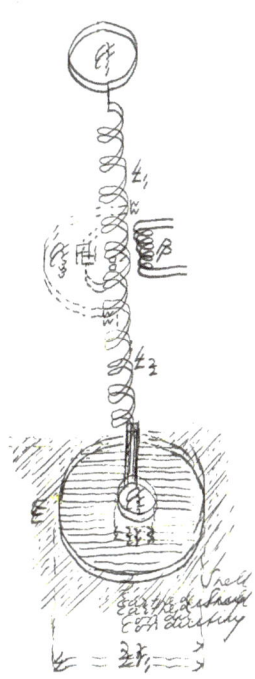

The driving system would probably not be directly grounded, but it would be capacitively coupled to the Earth. Something as in the sketch on the note from Jul. 23, 1900, doc. nr. 470.0250A, shown here to the left.

This system would discharge onto a movable sphere high up in the tower which was connected to the cupola. The cupola had a ground connection as is described in the Foreclosure Appeal proceedings (section 738):

This big ball on top of the tower, you could not tell what it was made out of, whether it was brass or steel, as the ends of the wires where it had been grounded had rusted out and blown away, and there was a thousand and one little wires sticking out in every direction, so you could not see what it was made up of.

The cupola and its ground connection would constitute the "free system":

---

10 300 HP ≡ 223.71 KW
11 400 HP ≡ 298.28 KW

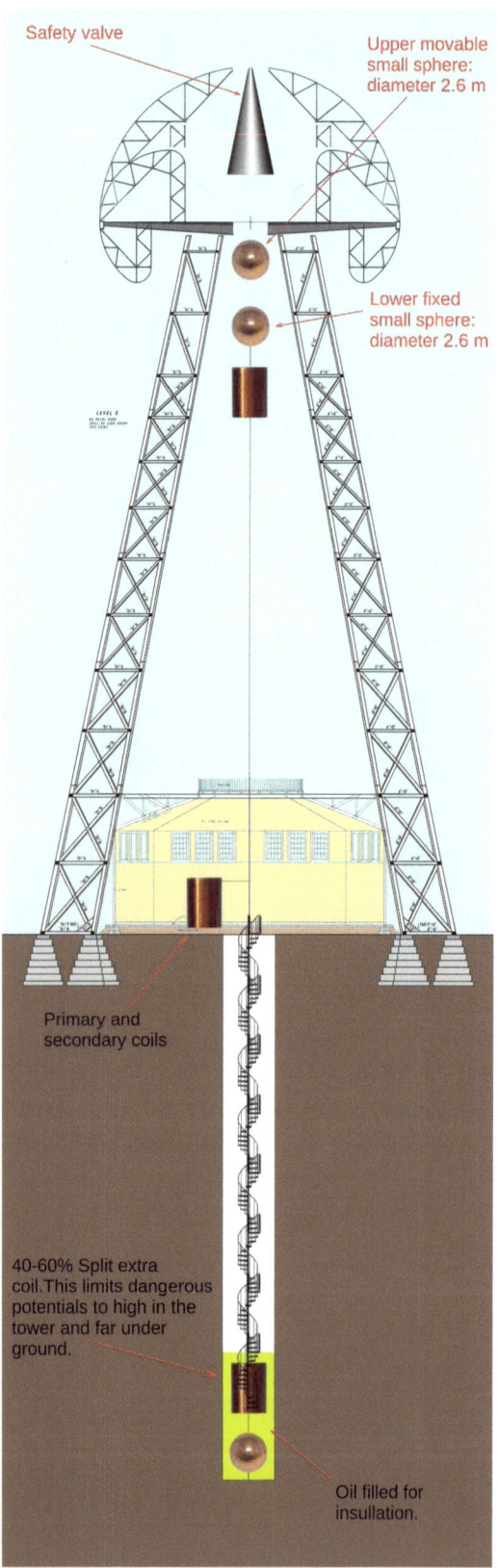

Safety valve

Upper movable small sphere: diameter 2.6 m

Lower fixed small sphere: diameter 2.6 m

Primary and secondary coils

40-60% Split extra coil. This limits dangerous potentials to high in the tower and far under ground.

Oil filled for insullation.

This next bit is speculation, but I think it is a perfect solution to otherwise very difficult problems. A half-wave resonator develops high potentials at its extremes, while in the middle the potential will be low. It would therefore make sense to cut this half-wave resonator into two pieces and put one below ground and one high up in the tower. The wire connecting them will always be at a relatively low potential. This idea, though again speculation, is supported by many facts. An important clue is found in a note (236.0096A) of June 26th, 1903 in which Tesla wrote that the average pressure on the conductor going down the shaft would be 10,000 Volts.

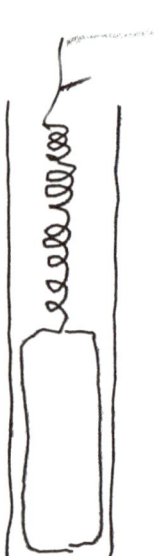

Another equally important clue we find in a note (470.0178A) dated Sept. 14th, 1903, that contains this sketch of an oil-filled well with a coil and a "top-load" inside.

Several other notes point in this direction, such as 470.0250A, which was published in "From Colorado Springs to Long Island" (see sketch on prev. page) and 470.0266A which shows a similar set-up.

Appendices

Museum Documents Referenced

Other Relevant Museum Documents

Box	Doc. nrs.	Description
335	001-005	About iron sheets in shaft below tower and water pump
354	002-007	Sun Earth calculations concerning gravity and electric forces
411	036-038	Ground resistance/glass plates condenser/
411	039-046	Ground resistance
411	047	Glass plates condenser
470	001-406	Notes from the Long Island Laboratory /technical notes/.
471	001-424	Notes from the Long Island Laboratory /technical notes/.
476	062-282	From the Long Island Laboratory /technical notes/.
477	030-054	Technical notes
478	001-013	Technical notes
478	021-041	Technical notes
508	001-068	Notes
508	383-410	Notes on distribution through the Earth
508	411-442	Various notes and drafts
508	443-444	Notes on covering cupola
538	309-320	Cost calculation to cover tower

Kenneth L. Corum on the cupola coverings

Tesla had plans for two types of perturbations for the top of the Long Island tower. First there was a design to increase the surface area of the charge terminal. This was a configuration of many bumps as shown in Tesla's illustration of the planned appearance of the Wardenclyffe Tower. It is my belief that Tesla planed that this would increase the corona around the charge terminal due to increase of the electric field at these small bumps. He comments about the corona on the tower in Colorado Springs. He saw this as a lot of free charge that could increase the capacity of the elevated terminal. It has been my experience that this is true but comes at a cost of lost power to maintain the corona in air. I am sure Tesla knew this and moved to the second arrangement of gas filled bottles on the charge terminal. In this case, the glass is not a dielectric but a way to keep a vacuum. This was much like his fluorescent tubes. The advantage of these bottles is that they were evacuated and a partial pressure maintained of a pure gas. There are many Tesla notes in the museum about this where he describes testing different gasses around 1905 (refers back to his Houston Street lab experiments) where he measures the ionization voltage required for gasses like hydrogen, nitrogen, argon, helium, and neon. (I remember demonstrating this for Dr. Marincic in the mid 80's. A single electrode bulb can give you quite a jolt if you touch it.) These bulbs were a way to reduce the great loss of corona and still maintain free charges (increased capacity).

I am sure Tesla wanted the bulbs but found it too costly and opted for the bumps. His plan may have been to replace the bumps with bulbs at a later time. Note that he does use bottles on his death ray tower.

Space reserved for notes
Space reserved for notes

QR Code Links to Websites

Tesla Science Center

Nikola Tesla Museum

Electric Spacecraft Journal

Order Blueprint Replica's

EMF Review - Kyle

E-Media Press

Energetic Forum

Wardenclyffe Research

3D scan Wardenclyffe

Youtube - Ernst Willem